# Basic Electronic Circuits
## Part- 4, Understanding Test Equipment

by Paul Honore' W6IAM

ISBN 978-1-300-75297-4

# Contents

## Foreword

I remember a cartoon-I think it was in *The New Yorker.* It showed a man and a small boy standing outside a store that was plastered with signs announcing "GOING OUT OF BUSINESS" and "FINAL SALE, EVERYTHING MUST GO." The man was saying to the boy, "Some day, my son, this will all be yours."

Part-3 of this series, *The Practical Radio Shack,* had just been released when a purchaser inquired, "You didn't say much about test equipment. I'd really like to know how to use an oscilloscope. I've always wanted to use one but I've been afraid to try."

So, it's "Deja-vu all over again!" Here's part-4 of my 3-part series. I promise to keep it simple so as not to scare anyone. I also promise never to promise again to not write any more on the subject.

## Introduction

Unlike machines that have pistons, valves, cams, gears and the like, whose operation can often be observed and adjusted without the aid of special instruments, electronic devices have no such indications of their operation. Electrons flow silently and unobserved, through wires, creating currents to produce desired actions such as a relay closure or an audible signal from a loudspeaker. If the relay fails to close or the loudspeaker fails to produce sound, or worse, smoke and fire emanate from the circuit, you must use external means to make the nature of the fault visible.

Instruments to test electronic circuits have been around for a very long time and most have not changed much since their invention. For instance, sometime around 1880, the German physicist, Heinrich Herz built the first high frequency signal generator. As soon as it became available, the French researcher, Arsene D'Arsonval, used it to study the action of alternating current on muscle contraction. To quantify his observations, D'Arsonval invented a sensitive meter to measure electric current. Both the signal generator and the D'Arsonval meter movement are in common use today.

Whether you are a professional, an amateur experimenter, or simply want to fix a simple electric device that's gone "on the fritz," you'll need some test equipment, be it an ohmmeter to check for circuit continuity, or a whole arsenal of instruments to perform analysis and alignment of a high frequency transceiver. What I propose, here, is to show you how these test instruments work; point out some of their benefits and limitations, and describe how to use them with a reasonable degree of confidence.

I must confess, I muddled through my early years as a ham radio experimenter without much in the way of test equipment at all. Back then, with the vacuum tubes and the dinosaurs, a lot could be done without any. To test an audio circuit, you could simply use a capacitor as a probe to inject 60 Hz hum, picked up by your body from nearby house wiring, to isolate a faulty amplifier stage. Continuity in a wire could be checked, using a battery and light bulb to complete a circuit. RF output from a transmitter could be checked by substituting a 100W incandescent lamp bulb for the antenna, or by holding a fluorescent lamp near the tank coil and tuning for maximum brightness. I could go on but it's pointless. Electronic equipment has become more complex and so have the instruments needed to test and align it.

As soon as I could, I acquired my first multimeter. It was a *Knight* kit. I not only learned something by building it, but it enabled me to experiment on a grander scale. It remained my only test instrument until I got a job at Stanford University's WW Hansen High Energy Physics Laboratory. It was there that I was exposed to my first oscilloscope -- a four channel *Tektronix* that looked like something out of Dr. Frankenstein's laboratory and it scared hell out of me. I learned how to use it and I was hooked. I had to have one. I didn't have the means to own a *Tektronix* so I built my own - another *Knight* kit. It was pretty mediocre compared to the *Tektronix*, but it expanded my abilities as an experimenter tenfold. I eventually did acquire a used *Tektronix*. It served me faithfully for more than 40 years, until one day it quit, never to work again. I now own a *Velleman* PC oscilloscope. It's not as versatile as my old Tektronix but it serves my current needs admirably.

Let's hope you are fortunate in being able to own the equipment you need. There are not as many kit instruments available as there once were, but the few that are out there are generally less expensive than those of 50 years ago, and they do more and do it better. If you're handy with a soldering iron, you can assemble a respectable collection of test instruments for a modest investment.

## Unit-01, Analog Meters

Fig 1-1

In 1820, the Dane, Hans Christian Oersted, discovered that an electric current, flowing trough a wire, would deflect a magnetized needle. This led to the development of the galvanometer, named for the Italian biomedical researcher, Luigi Galvani. The galvanometer is considered to be the first electronic test instrument and it is still in use in laboratories, today.

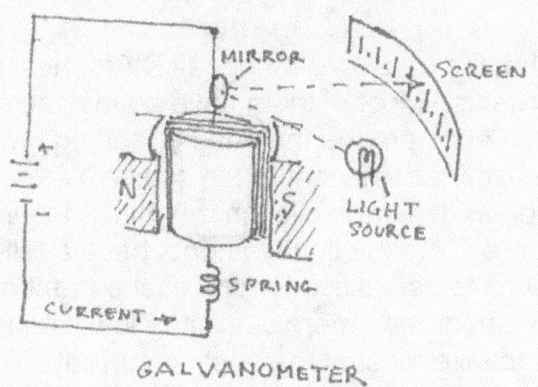

Fig 1-2

The galvanometer is a moving coil and mirror, suspended between fixed magnet poles. When an electric current is passed through the coil, the coil rotates, deflected by the magnetic field, against the pressure of a light spring, The amount of rotation is in direct proportion to the applied current. A light beam is deflected by the mirror onto a graduated screen to provide a visual indication.

## D'Arsonval movement

A French researcher of the 19th Century, Arsene D'Arsonval, was interested in the effects of electrical stimulus on muscle contraction. Galvanometers in use at the time were not sensitive enough for his experiments, so he set about to improve their performance. Ultimately, he designed a variation of the instrument known as the D'Arsonval movement - the analog meter movement still in common use. What D'Arsonval did was to improve the bearing system and replace the mirror and light source with a needle that would "sweep" across a graduated screen to make a robust portable meter.

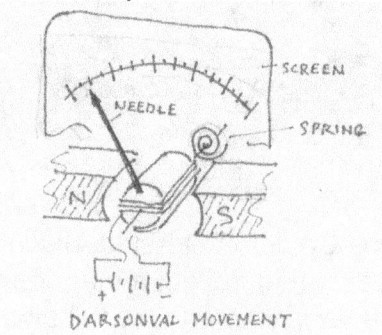

Fig 1-3

## DC Ammeter

Regardless of what they are measuring, all D'Arsonval movements are current-reading instruments. Most of them have a small screwdriver adjustment to set the spring tension and zero the the needle reading. More expensive meters include a strip of mirror adjacent to the scale. When viewing the reading, you can align the needle with its reflection in the mirror to eliminate parallax error. Without any modification, D'Arsonval movements have a full-scale deflection of a few microamperes,-much too sensitive for practical current measurements.

To make the movement read more current, a low resistance shunt is placed across the movement to take most of the applied current. You'll remember from the first book in this series, *Analog Circuits*, that if two or more resistors are connected in parallel, current is divided between them in inverse proportion to their value.

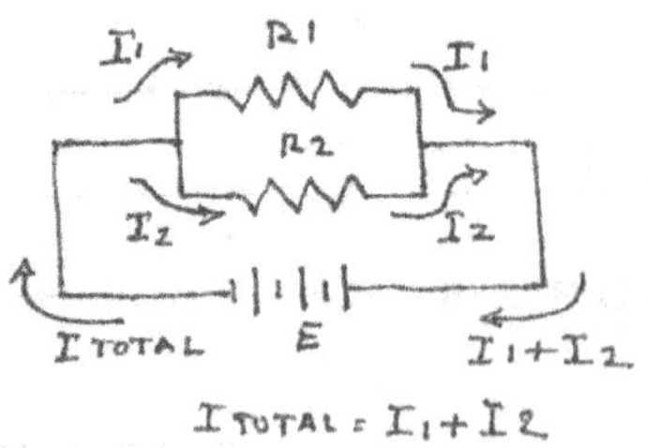

Fig 1-4

The meter coil constitutes one resistor in our parallel circuit and the shunt is the other resistor.

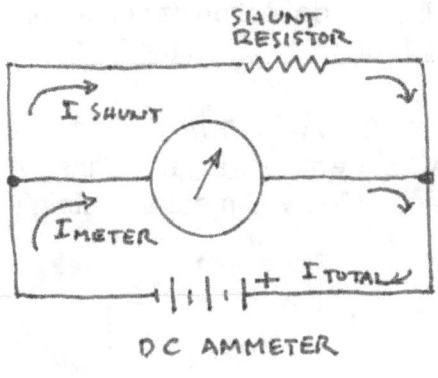

Fig 1-5

The shunt will take the bulk of the current and the meter movement will take its normal safe current to deflect the needle. It used to be common practice to calculate the shunt value, knowing the resistance of the meter movement and the applied voltage. For instance, suppose you have a meter that has an internal resistance of 10KΩ and requires 10mA to deflect full scale. What shunt resistance would you need to measure a current of 10A?

Answer: You can solve by ohm's law. Therefore the applied voltage it takes to deflect the meter full-scale is E=!R, E=.010X10,000, E=100V

Since the shunt must take all but 10mA of current it must take 9-.010A, = 8.99A. Solving by ohm's law, we get R=E/I, R=100/8.99, R=11.12Ω for the shunt value

Modern practice dictates the use of standard shunt values that produce known voltage drops when current is passed through them. These are available commercially at modest cost. They are designed to air-cool for temperature stability.

Fig 1-6

The shunt in Fig 1-6 is rated such that 100A of current will generate a 50mV drop across its resistance. When the meter and shunt are connected in parallel, a meter with a full scale deflection of 50mA will deflect full-scale when 100A of current is passed through the circuit

## AC Ammeter

If you want to measure AC current, you'll have to insert a diode in series with the meter to rectify the voltage. Most AC meters contain internal diode rectifiers.8

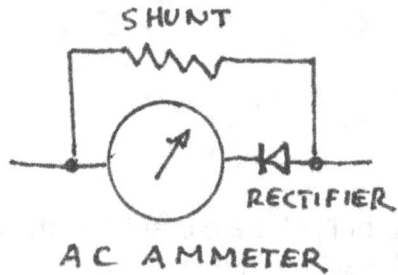

Fig 1-7

For a permanent installation, it is customary to insert a current transformer in the circuit to be measured. This can be done in one of two ways. Either the primary transformer winding can be placed in series with the circuit, or a current-carrying wire can be passed through a toroidal transformer core. The alternating magnetic field that surrounds the wire will be coupled to the transformer coil and read directly by the meter.

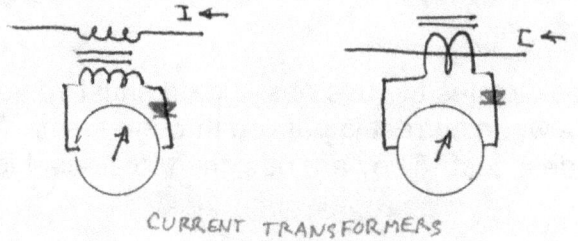

Fig 1-8

Here's an example of a toroidal current transformer.

Fig 1-9

To use it, simply pass a current-carrying wire from the circuit to be monitored through the hole in the toroid and connect the meter to the terminals at the top9

*Clamp-on Ammeter*

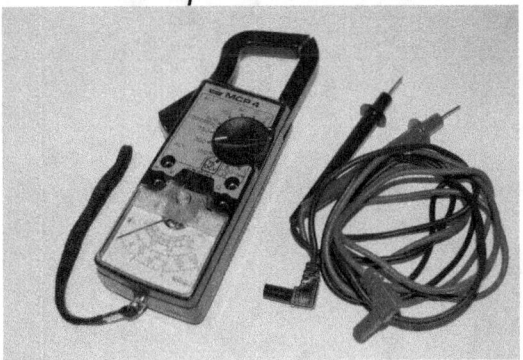

Fig 1-10

To make temporary measurements of AC current, a hand-held clamp-on ammeter is useful.  It is a portable variation of the toroidal transformer and meter combination.  The current transformer is mounted internally and the metal transformer core is "wrapped" around the wire carrying the current to be measured.  The core is split and hinged so that it can be placed over the wire without interrupting the circuit to make a connection.

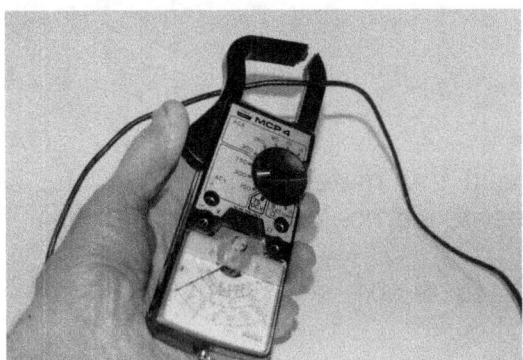

Fig 1-11

This model has some bonus features.  Using the supplied test leads, it can also measure up to 750VAC, 75VDC, and with the ohmmeter scale, it can be used to measure circuit continuity.  Altogether, it's a very useful instrument for the electrician.

## *Voltmeter*

Once again, the D'Arsonval; movement forms the basis for most voltage measurements.  All that is needed is to insert a resistor in series with the movement to determine full-scale voltage deflection.

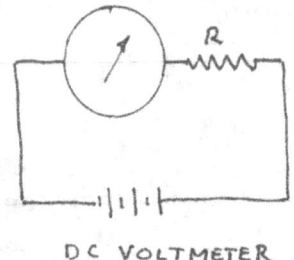

DC VOLTMETER

Fig 1-12

As with the AC ammeter, if you want to measure AC voltage, you'll have to insert a diode rectifier in series with the meter.

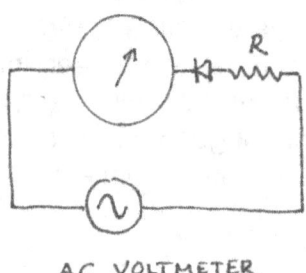

AC VOLTMETER

Fig 1-13

Suppose you have a 1mA meter and you want a full-scale deflection of 1000V.  What series resistance would be required?

Answer:  R=E/I, R=1000/.001, R= 1000000$\Omega$ (1M$\Omega$)

Regardless of how it is used, the  meter movement will draw some amount of current to deflect the needle  The customary way of expressing the current drain, and therefore the load the meter will place on the circuit being measured, is in Ohms per Volt.  If you are measuring a 100V source and using a20,000 ohms per volt meter movement, you will put a load of I=E/R, I=100/20,000, I= 0.005A. on the circuit you are measuring.  If you are using a 1000 Ohms per volt movement, your meter will draw 100/1000 =  0.1A.  This puts us in the realm of subjective determination.  If you just want to see if there's power on a circuit, a low ohm-per-volt meter will do.

If it is likely to put enough load on the circuit to affect its operation, you need a better instrument.  For most measurements, 20,000 ohms per volt is sufficient. If even that imposes too much load, you'll need a more sophisticated approach.

## *Vacuum Tube Voltmeter*

Fig 1-14

50 years ago, the standard way to avoid overloading sensitive circuits when measuring them, was to use a vacuum tube voltmeter (VTVM).  The use of a vacuum tube amplifier allowed a very high input impedance (several megohms), making for a negligible load on the circuit being tested.  The downside of such instruments was their tendency to drift as the vacuum tubes warmed up.  Also, since they required high voltages to operate the vacuum tubes, they could not be considered portable since the power supply was usually operated from 110VAC mains power.  VTVMs are obsolete now but there are still some used VTVMs available from sources on the internet.  For reference, here's the schematic for my own, ancient *Heathkit* VTVM

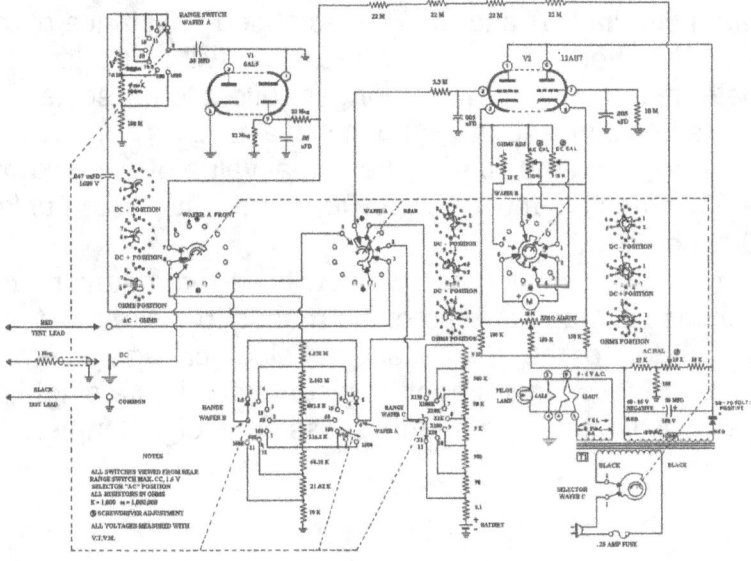

Fig 1-15

## *Ohmmeter*

So far, we've seen how current and voltage is measured. Now, we'll see how the same meter movement is used to measure resistance.

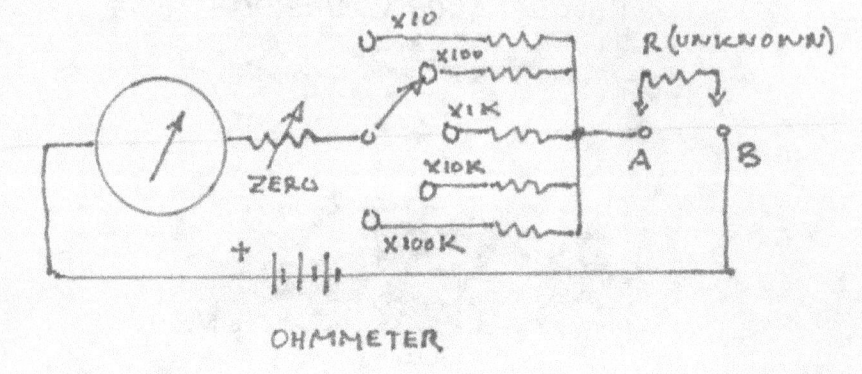

OHMMETER

Fig 1-16

In this application, the meter movement is placed in series with a DC power source and a known resistor value (Multiplier). To use it, a short circuit is applied across terminals A and B, causing the meter to deflect to full scale. Precise zeroing is done with a potentiometer. The short circuit is removed and a resistor of unknown value is inserted between terminals A and B. It's value will be indicated on a scale graduated in Ohms. You would multiply the reading by the selected multiplier scale.

Let's take an example. Suppose the meter is selected to the X100 multiplier scale. It is "zeroed" at full scale deflection, and a resistor of unknown value is inserted. between terminals A and B. The increased resistance of the series circuit, due to the insertion of the unknown resistor, will cause less current to flow through the meter movement. It will no longer deflect to full scale, but will deflect to a mark on the scale that is to the <u>left</u> of full-scale. Let's suppose the needle settles on the 500 mark on the scale. The actual value of the unknown resistor would then be 500 times the multiplier (X100). The value of the unknown resistor would be 50,000 ohms (50K$\Omega$).

Using the D'Arsonval movement, we now have the means to measure all three parts of Ohms Law, and therefore, the means to make most of the measurements needed for the maintenance of electrical equipment. In Unit-3, I'll show how all three of these instruments are combined into a single "multimeter" but first, in Unit-2, I'll digress for a short discussion of digital meters.

## Terms to remember

**Current transformer**    Transformer in series with an AC circuit to couple the induced magnetic field to an ammeter for measurement.

**D'Arsonval movement**    Analog currenet measuring movement comprising a moving coil between fixed magnetic poles

**Galvanometer**    Laboratory instrument for measuring minute electric currents.

**Ohmmeter**    Device to measure the value of an unknown resistor by measuring current in a series circuit

**Shunt**    Low value resistor in parallel with a meter movement to increase its current reading ability

**VTVM**    Vacuum tube voltmeter

## Formulas

**Ohms Law**    $I=E/R,\ R=E/I,\ E=IR$

## Unit-01 Exercises

01-1)    A D'Arsonval movement has a full scale deflection of 50mA at 10V.  What is the resistance of its moving coil?

01-2)    What shunt resistance will be needed for it to read 1ADC full scale?

01-3)    What series resistance would be needed to make it read 100VDC full scale?

01-4)    An unknown resistor reads 24 on the X10 Ohmmeter scale. What is its value?

## Unit-2, Digital Meters

Fig 2-1

Today, all food has to be "organic", and everything with an alpha-numerical readout must be "digital". Let's be clear from the beginning. Just because a device has am alpha-numeric readout instead of a needle and graded scale, doesn't mean it is digital. You can purchase a five dollar voltmeter with an alpha-numeric readout but is a low grade analog instrument. The display is digital in the sense it is alpha-numeric, but A/D conversion is only for the display, not for the value being measured. mmoreover, no interpolation between whole numbers is possible so the accuracy of such a device is worse than for an equivalent analog meter. The schematic for this type of hybrid meter looks something like this:

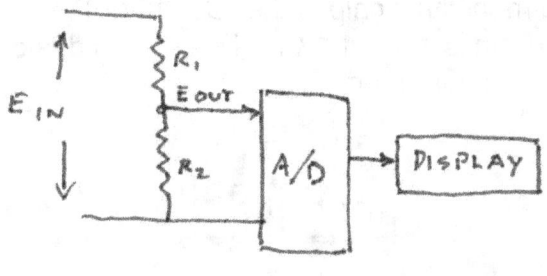

Fig 2-2

R1 and R2 form a voltage divider . R1 is adjusted to produce a full-scale value across R2. when the desired maximum voltage is applied to the input. The voltage across R2 is converted to digital display format by a monolithic integrated circuit chip to actuate the display. The IC and alphanumerical display chips are available off the shelf at such a low cost that it no longer makes sense to use a D-Arsonval movement for most applications.

By comparison, a true digital instrument looks more like this.

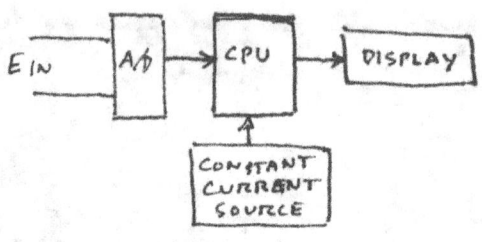

Fig 2-3

Notice that the input value is immediately converted to digital format and passed to a microprocessor for calculation before being displayed. The use of a microprocessor chip allows some interesting operations to be performed, such as a floating decimal point, making the device more accurate and more versatile. I'll discuss this in more detail in Unit-3.

A typical high performance digital meter is built around a complex monolithic circuit that includes an A/D converter, clock multivibrator, display drivers for an LED display, and a central processing unit (CPU). I described the individual circuits in the second book of this series, *Digital Circuits*, and will not repeat the information here. All you need to know is that it's all contained in one chip. Trust me.

Ussing this technology, few external components, other than a 5V DC power supply and an LED display are needed to build any kind of meter desired. An example of such a monolithic chip is the ICL7107. Here's the way it appears from the top. The diagram is aabout 4X real size. It is designed to be flow-soldered onto a printed circuit board.

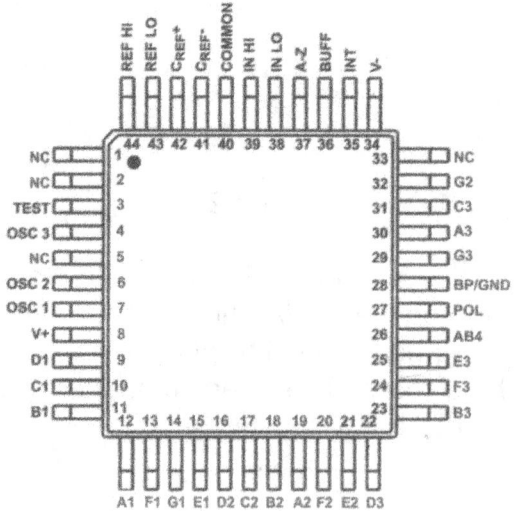

Fig 2-4

Here's a schematic for a digital DC voltmeter (DVM), using the ILC7107. R4 is chosen to scale the meter for the desired maximum voltage display.

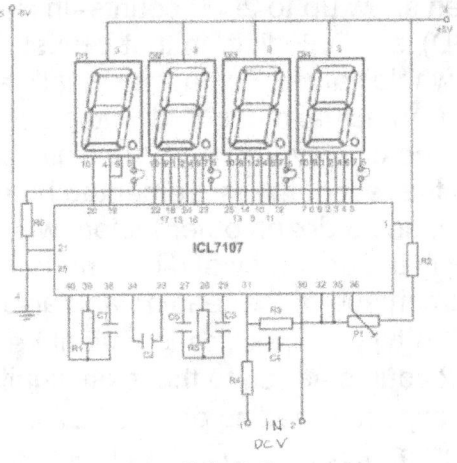

fIG 2-5

A slight modification to the input circuit allows the same device to read DC Amperes.

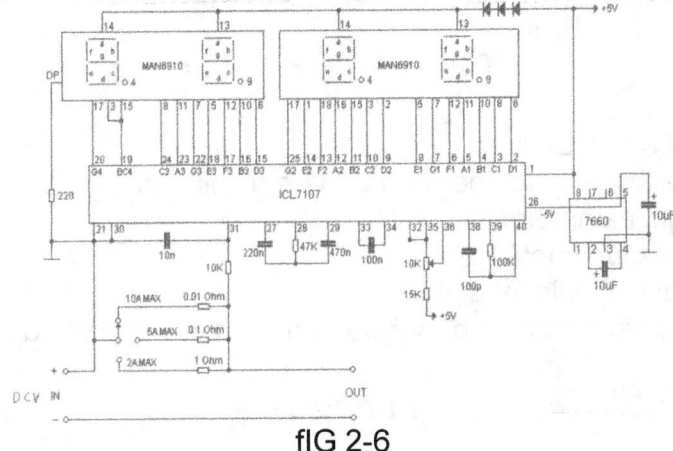

fIG 2-6

Having said all that, what about accuracy?  It's tempting to assume a digital meter is more accurate than an analog meter.  Perhaps.  The fact is, a digital meter is only as accurate as the that of the components it contains and on such external factors as temperature, humidity, and noise, (both internal and external). and on the stability of internal references.  Any or all of these factors can affect the accuracy of the measurement significantly.

Let's take the display, itself. Displays are rated in counts, (the number of digits they can display). A display that shows a maximum of four digits is called a 3-1/2 digit display and can show up to 2000 counts- in real numbers, 1999. The least significant digit (LSD) is a "half digit" in that it must decide to increase or decrease to the nearest whole value. For instance, let's assume the meter to be a 100V, 3-1/2 digit scale. An actual measured voltage of 5.5V will be displayed as 5.500. The same voltage displayed on a 4-1/2 digit scale would read 5.5000. Of course, that assumes the rest of the instrument to be perfect. In the real world, noise, temperature, and a dozen other factors will alter the final reading.

Manufacturers specify the accuracy of digital meters in two ways. They specify accuracy as a percentage of the reading and also as an offset amount for the least significant digit. A typical specification might be ±5%+2. That is, (±5% of the measured value +2 counts added to the least significant digit.) So, on our 3-1/2 digit scale, we can expect a reading of 5.5±0.275, (+2 counts added to the LSD). The result will be from 5.227 to 5.777. For casual measurements, this might not be a significant error but if you are calibrating a laboratory instrument, it could be significant indeed!

In this unit, I've described the basic workings of a digital panel meter and shown some of it's pitfalls. In the next unit, I'll discuss the multimeter.

## *Terms to remember*

| | |
|---|---|
| A/D | Analog to Digital |
| Count | The highest number that can be displayed |
| DAM | Digital ammeter |
| DVM | Digital voltmeter |
| LSD | Least significant digit |
| LSD count | Number added to or subtracted from the displayed LSD |

## *Formulas*

Accuracy = %of measured value ±number of counts added to the LSD

## *Unit-2 exercises*

02-1)　A 2000 count meter with a 100V maximum is used to measure 25VDC. What would the display show?

02-2)　The manufacturer of a digital voltmeter claims an accuracy of "1% of the indicated reading +3". Assuming an actual voltage of 25.5VDC, what maximum and minimum indications can be expected?

## *Unit 3, Multimeters*

Fig 3-1

In Unit-1, I showed how a D'Asonval movement could be used to measure Voltage, Current, and Resistance. Panel meters for these measurements are useful for permanent installations but, early on, it became customary to incorporate these measurements into a single hand-held instrument. The Volt-ohm-miliammeter (VOM), or Multimeter, has remained the basic tool in the technician's traveling kit ever since. It's a given that you should own one or more of these handy instruments but which one? Of course the ultimate decision is yours to make, but I'll try to strip some of the mystery from the bewildering array of choices and help you to decide wisely.

To confuse matters, both analog and digital VOMs are called multimeters. Analog versions are referred to by the acronym VOM. Their digital counterparts are referred to as DMMs. Regardless of what you choose to call them, or how many different functions they perform, all of these meters operate in the same way. They use a multi-position switch to select various shunts and calibrating resistors, in combination with an internal battery, to determine the type and scale of measurement,

### *.Analog Multimeters*

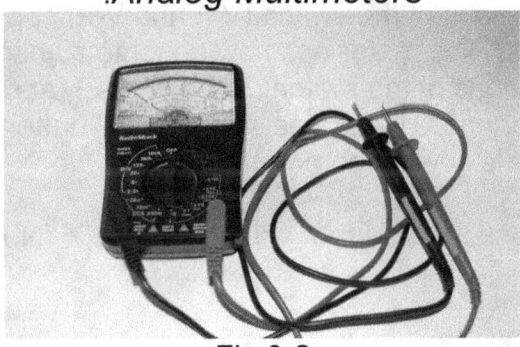

Fig 3-2

In its most basic form, the schematic for an analog multimeter looks like this:

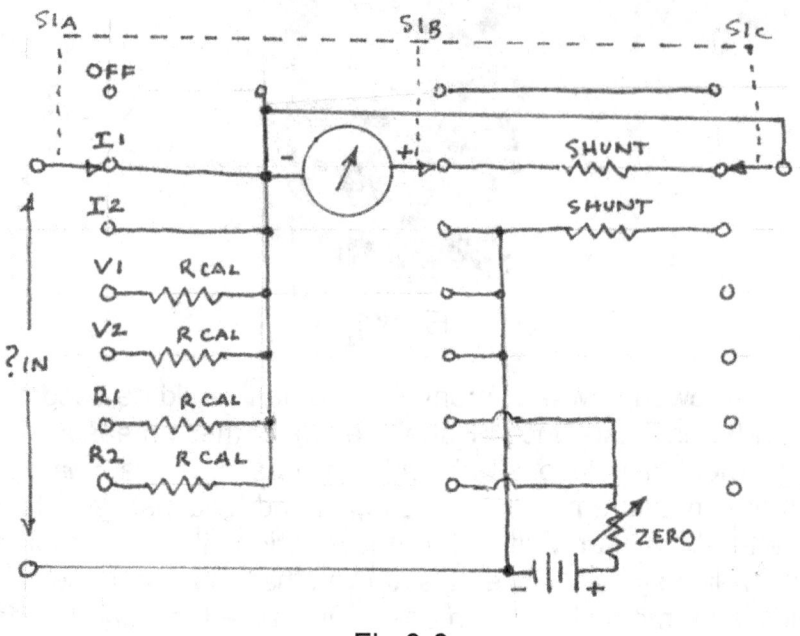

Fig 3-3

A pair of test leads connect the circuit to be tested to the terminals at the left. Starting from top and working down, In the OFF position, switch S1(B and C) puts a short circuit across the meter to "Dampen" the moving coil with regard to the magnetic field and prevent it from moving as the meter is roughly handled during transport.

In the next two positions (I-1 and I-2), shunt resistors are applied across the meter to allow current measurements. The meter is now an milliammeter/ammeter.

The next two positions of the switch (V-1 and V-2) apply scaling resistors for voltage measurements, The meter has now become a voltmeter.

The last two positions of the switch (R-1 and R-2), apply calibrating resistors in series with the meter and S1(B) selects an internal battery to measure resistance. The meter is now an ohmmeter. Handy, wot?

Now, to the real world. Here's the actual schematic for the VOM in Fig 3-1

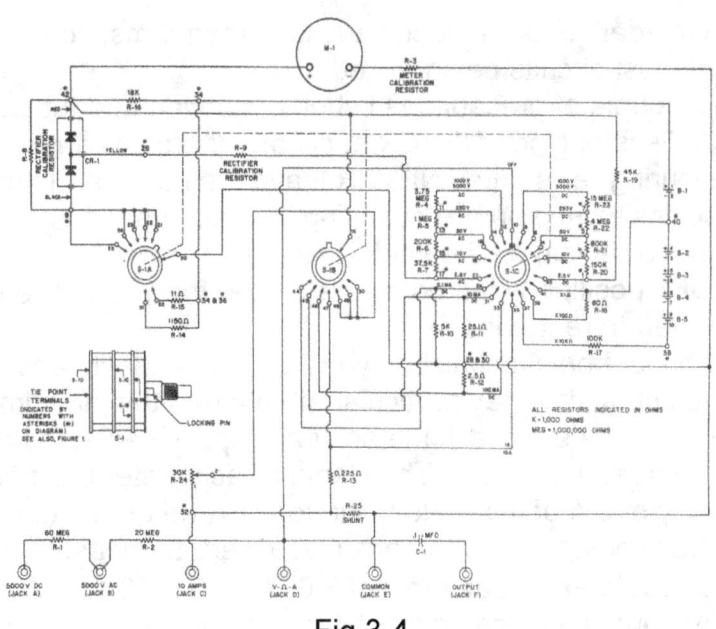

Fig 3-4

Don't let it scare you. It looks a lot more complicated than the simple schematic in Fig-3 but I assure you, it's pretty much the same. Diode rectifiers have been added to measure AC voltages and there are more switch positions, making for more measurements. Here's a list of the maximum values it will read for the various switch positions.

2.5, 10, 50, 250, 1000 and 5000VDC
2.5, 10, 50, 250, 1000 and 5000VAC
10 mA, 100mA, 1A and 10ADC
RX1,RX100 and RX10kΩ.

Pretty impressive, huh? It's a 60 year old *Knight* kit multimeter.-- the first test instrument I ever owned. I don't use it any more but I still have it. It still works! Now, let's take a look at the meter in Fig 3.2 It's a Radio Shack meter costing about $30. It is a good example of a modern analog meter with an impedance of 20,000 ohms per volt DC and 9,000 ohms per volt AC. It has 19 ranges with the following maximum values

2.5, 5, 25, 125, 500 and 1000VDC
10, 50, 250 and 1000VAC
50μA, 25mA and 250mADC
RX1, RX10, RX100, RX1k and RX10kΩ
Audible beeper for continuity checking

The input impedance for the test leads is 20megohms, so it will not cause undue loading of most circuits being tested.

There are a couple of precautions I should mention and they apply to any multimeter, regardless of type. NEVER apply an external voltage to the test leads when measuring resistance. NEVER leave the meter switched to one of the ohmmeter scales when you are not using it. Remove the internal battery if the meter is going to remain on the shelf for a long period of time. ALWAYS switch it to the OFF position when not in use. Over the years, I've violated all of these rules and lived to regret it.

If you are an electronic technician, you'll want a versatile and accurate multimeter. When choosing, pay particular attention to the input impedance. The higher, the better. Also, check the accuracy specification to be sure it will meet your minimum requirements. The Radio Shack meter I just described claims an accuracy of 3% of full scale value for DC and ohms readings, and 4% of full scale for AC readings. That means if you were to measure 500VDC on the 1000V scale, you could expect an error of $\pm.03 \times 1000 = \pm 30V$ When you apply that to your 500V source, you can expect it to read anywhere between 470 and 530V. If that margin of error is acceptable, well and good. If not, you might want to consider shopping around for a better instrument.

Price is not necessarily an indication of quality. I've seen meters costing many hundreds of dollars whose accuracy is no better than a 25 dollar model. And, while we're on the subject, you might want to purchase one or two of the really cheap (five dollar or so) multimeters available at most warehouse stores and auto supply shops. They are useful when "banging around the shop, checking for voltage and continuity in an automobile or the house wiring, and if you do any tower climbing, to work on antennas for instance, you won't feel bad when you drop the five dollar meter. from a height of 70 feet or so.

## Digital Multimeters

Fig 3-5

Except for the alpha-numeric display, digital and an analog multimeters appear to be about the same. It's true that they are similar in function and appearance but they are about as different as can be in the way they go about their jobs. Here's a simplified block diagram of a digital multimeter.

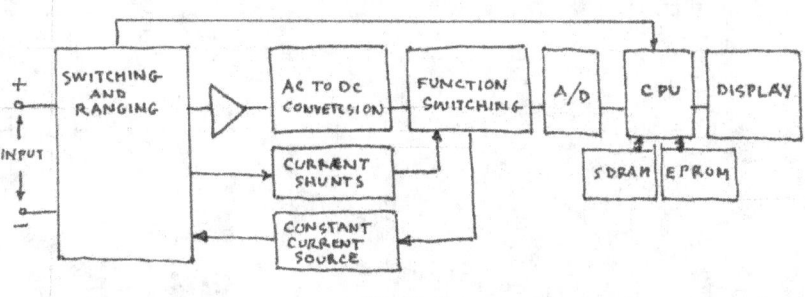

Fig 3-6

The heart of the DMM is a CPU chip that controls all functions, calculates the measurements, and passes the output to an alpha-numeric display. This opens new possibilities. For instance, many DMMs automatically detect polarity of the input signal and make necessary corrections. They may also detect the level of the input and automatically set the range to the proper scale for measurement. Here's a schematic for a relatively inexpensive hand-held DMM.

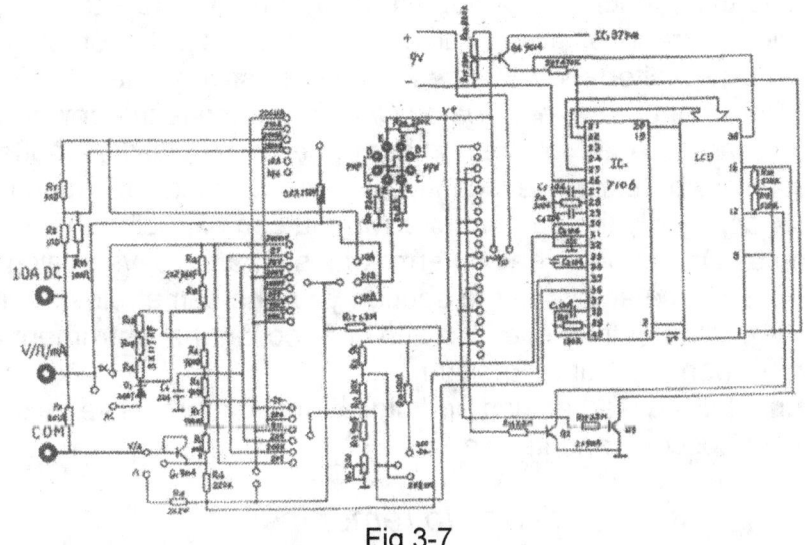

Fig 3-7

Notice that there are only two microchips - a CPU and a display chip. Except for a manual input switch, there are only a handful of resistors and transistors. This model even offers a panel socket to test transistors. It is a commercial DMM comparable to an analog meter in the $30 price range. By comparison, here's a schematic for a more complex, (although not necessarily more expensive) DMM

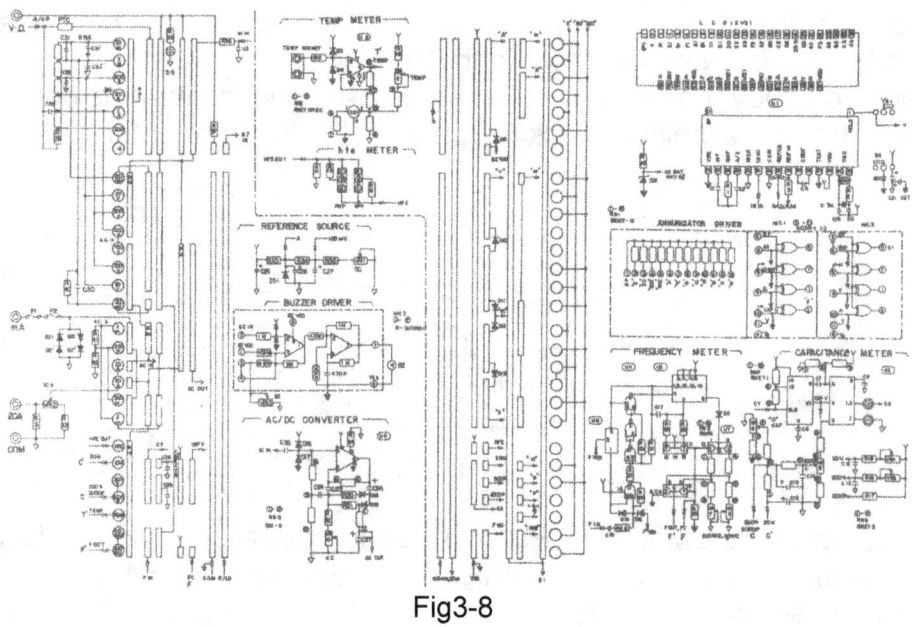

Fig3-8

This DMM offers full automatic ranging and polarity sensing. As a bonus, it can measure capacitance, temperature, and AC frequency. Capacitance measurement is accomplished by simulating an LC reactive circuit in the CPU and substituting the capacitor to be tested as an unknown reactance. The CPU then checks down a table of stored values and finds the value of Lz that nulls the reactance of the capacitor and produces a pure resistive value. The method was described in Part-1 of this series, *Analog Circuits*, Such is the power of a CPU!

Don't purchase a DMM for the number of bells and whistles it offers alone, although these may be attractive incentives. Pay particular attention to the specifications, especially the amount of error you can expect in the displayed data. In general, DMMs offer reading error of less than 1%, vs analog meters at greater than 3%. Once again, let me caution you, the same rules apply to DMMs as to VOMs regarding battery shelf life and connecting the instrument to a voltage source when measuring resistance.

In this unit, I discussed analog and digital multimeters. The next unit will describe the oscilloscope and its use

## *Terms to remember*

DMM        Digital multimeter
VOM        Analog multimeter

## Unit-04, Oscilloscopes

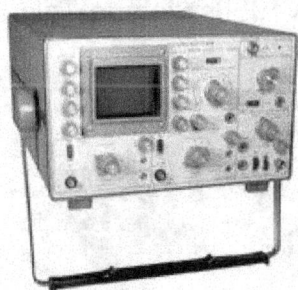

Fig 4-1

Early on, it became desirable to record waveforms for analysis. Chart recorders that used a moving pen to ink a waveform onto a traveling strip of paper were among the first to be used and they are still in common use today. Their ability to record rapidly changing events is limited, however, by the ability of the pen to respond fast enough. To compound the problem, a pen, operated by a moving coil, has a ballistic component that causes it to overshoot the mark and distort the recorded data. Therefore, chart recorders are limited to very slowly changing, or long term events, such as voltage changes over a period of hours or days

Mirror galvanometers, adapted to deflect a light beam onto a moving strip of negative film, gave improved response but they were still limited to about 10KHz response time. That's where the art of recorded waveforms stood until the development of the cathode ray tube (CRT) by Karl Braun in 1897. "Braun tube" oscilloscopes, as they were known then, remained a laboratory curiosity until sometime in the 1930s, when *General Radio* built the first portable oscilloscope, (In today's parlance, O-scope.)

## CRT oscilloscopes

Fig 4-2

A CRT oscilloscope uses a specialized vacuum tube, incorporating a high voltage electron gun that emits a beam of high energy electrons

The electron beam bombards a phosphor screen to produce a spot of light. Here's a CRT extracted from a 1950s vintage *Tektronix* oscilloscope.

Fig 4-3

Now, let's take a look inside to see how the thing works.

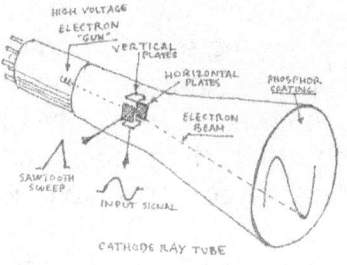

Fig 4-4

.  A sawtooth voltage is applied to a pair of horizontal plates to electrostatically deflect the beam.  Sawtooth generators were discussed in part-1, *Analog Circuits*, of this series,  As the voltage on the right-hand plate increases, the electron beam is deflected to move from left to right across the screen.  The phosphor glows at the point where the electron beam strikes it and The phosphor persistence keeps it glowing for a time after the beam has moved on. leaving a visible trace in the form of a bright line on the screen.

If a signal is applied to a pair of vertical plates, the electron beam is deflected vertically as it travels, so that it traces a visual picture of the waveform. When the sawtooth voltage reaches maximum, (and the electron beam is deflected to the end of its travel), it quickly sweeps back to its starting point. During the return sweep, he beam is "blanked out" to prevent the retrace from being seen.

In early oscilloscopes, the image was only stable for a single sweep across the screen because there was no way to synchronize the start time of the sweep with the event being observed, the next sweep would produce a waveform in a different position from the first, making it appear to travel as it was repeatedly sampled and displayed by the CRT.

Fig 4-4 illustrates the asynchronous relationship between the sawtooth voltage and the waveform being sampled.

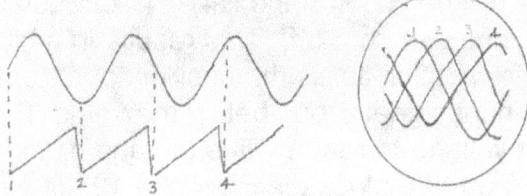

Fig 4-5

The solution to this dilemma came with the invention of a triggering circuit by Otto Schmitt in 1934. The Schmitt trigger allows the CRT trace to begin at the same point on the waveform each time it is sampled. . Let's assume the waveform to be studied is a sine wave. The Schmitt trigger can be adjusted to turn on at a specified voltage corresponding to a point along the rising sine wave (A), initiating a sawtooth voltage sweep each time the input signal reaches point (A) on the waveform. The result is a stable image on the screen.

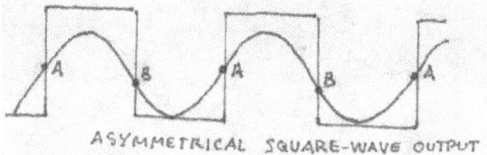

Fig 4-6

Notice that the trigger signal shuts down at a lower voltage level (B). This prevents noise at the input from generating a false trigger. Here's a schematic for a Schmitt trigger circuit. Of course, when the circuit was invented, vacuum tubes were used. In this illustration, I've shown the more modern, transistorized version.

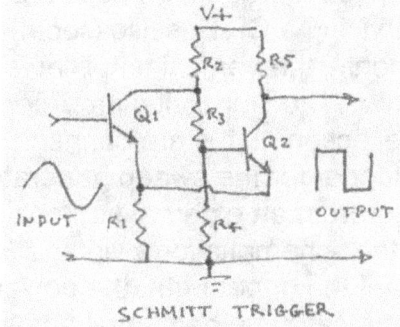

Fig 4-7

If there is no signal present at the input, Q1 is off and Q2 is conducting, It's output will be LOW. When the input signal rises above the threshold voltage (A), determined by voltage divider R2, R3 and R4, Q1 conducts, lowering the voltage at the base of Q2, causing it to shut off. This results in a HIGH output at Q2s collector. The output from Q2 initiates the oscilloscope's sawtooth sweep generator. When the input signal drops below the value (B), the circuit "resets" and waits for the input voltage to reach value (A) again, initiating another sweep.

In its most basic form, the block diagram of a 1940s vintage oscilloscope looked something like this.

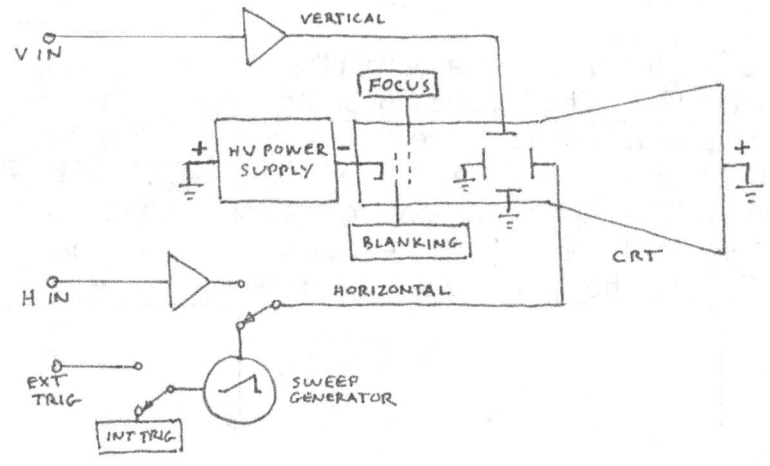

Fig 4-8

The electron gun portion of the CRT resembles a conventional vacuum tube in function. A cathode to emit electrons. The free electrons are focused into a beam which is attracted to a positively charged phosphorescent screen that serves as the anode. For safety reasons, because the cathode-to-screen voltage is usually in the order of several thousand volts, the screen is grounded and the cathode is operated at a negative potential. (That means all the components associated with the electron gun are at a high negative voltage. Exercise EXTREME CARE when servicing a CRT oscilloscope!).

For normal measurements, the vertical amplifier is connected directly to the vertical CRT deflection plates. The Horizontal amplifier is switched out of the circuit and the horizontal deflection plates are connected to a sawtooth sweep generator. Synchronizing trigger for the sweep generator can be supplied from the internal Schmitt trigger or from an external source.

For reference, here's the schematic for a vintage *Heathkit* oscilloscope of the 1950s era. The CRT is at the upper right, the power supply at the lower right, the vertical amplifier at the top, and the horizontal amplifier at the bottom. Controls and the sweep and blanking circuits are in the center.

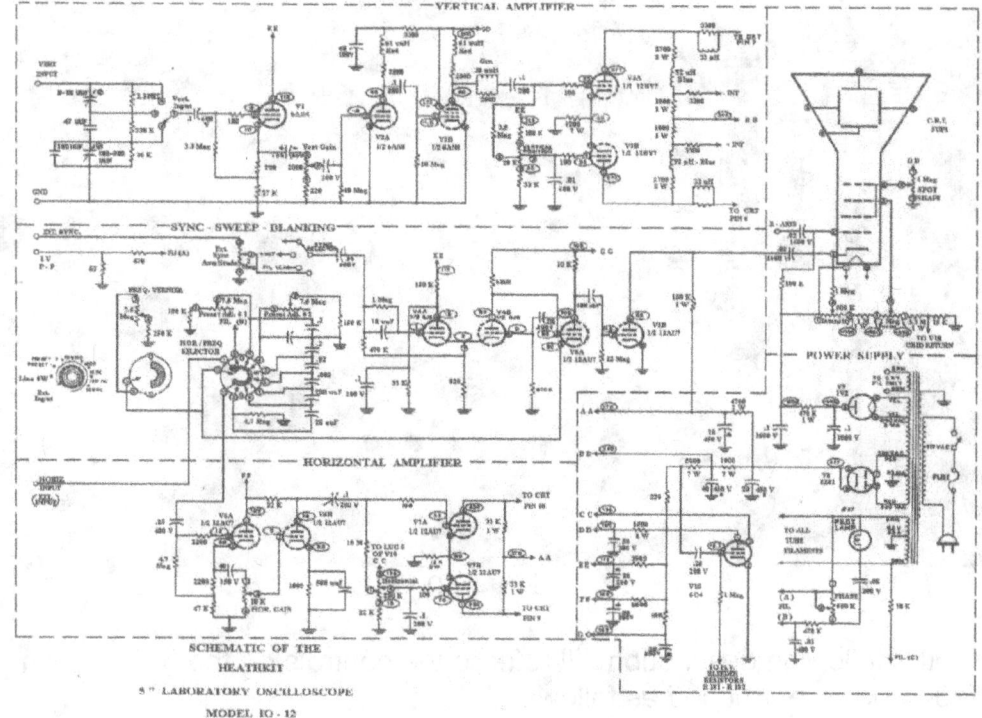

Fig 4-9

   Now, let's take a look at a CRT oscilloscope to see how to is used as a measuring instrument.  Since Tektronix has become the world standard for precision  oscilloscopes,  I'll use their 503 model as an example.  Not all scopes have the many features offered by Tektronix but the important functions are present, no matter what scope you are using.

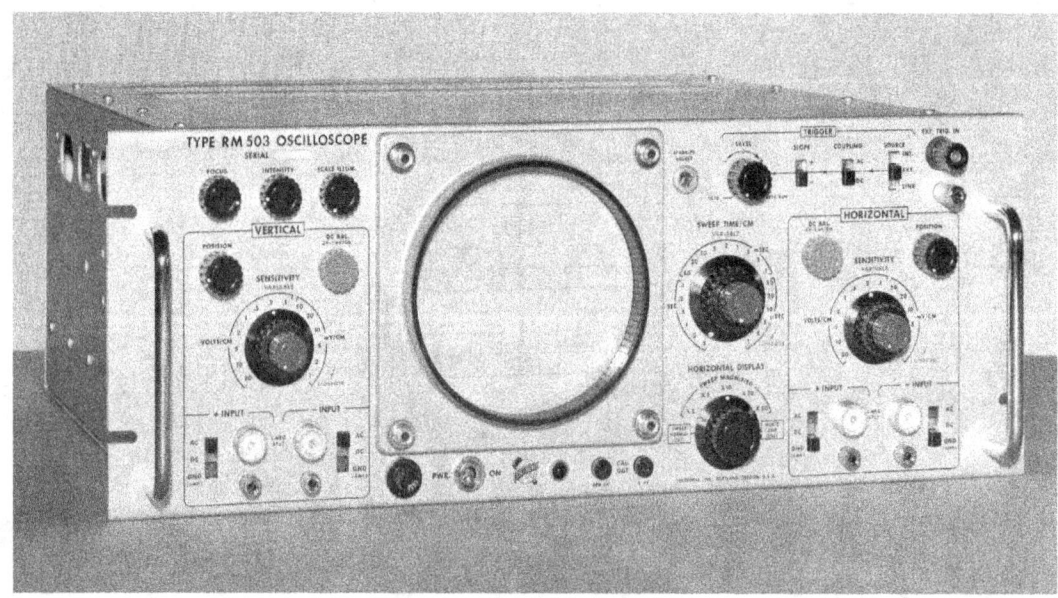

Fig 4-10

In the following description, I'll refer to the controls on the 503, shown in Fig4-10. They are grouped as follows.

**Upper left**- Beam focus and intensity, and gratical illumination.
**Center left**- Vertical amplifier controls
**Lower left**- Vertical input jacks and switches
**Upper center** CRT with gratical and sweep controls
**Lower center**- Power ON and calibration jacks
**Upper right**- Trigger controls
**Center right**- Horizontal amplifier controls
**Lower right** Horizontal input jacks and switches
**Bottom center** Power ON switch, pilot light, calibration jacks

Before powering up a CRT oscilloscope. turn the INTENSITY control to MINIMUM This is important. A very bright trace or an electron beam that is allowed to remain static for even a few seconds, can burn the phosphor and leave a permanent mark. Always start with the beam at minimum intensity and increase the brightness just enough to provide a readable image.

The CRT has a plastic gratical over its face. The graticle is divided into centimeters, both vertically and horizontally. This is to make it possible to dirctly read the amplitude and frequency of a waveform. More about this later.

Your scope should also have a SCALE ILLUMINATION control. On the 503, it is located third from the left in the upper left hand corner. It controls lamp bulbs at the edge of the gratical to make the lines visible.

There is also a FOCUS control in the same row of knobs.  It is used to focus the electron beam to a sharp  point on the screen.

Now to the important stuff.  There are separate VERTICAL and HORIZONTAL amplifiers.  Their controls are identical, so I'll only describe the VERTICAL amplifier controls.  The 503 has two inputs for each amplifier but most scopes only have one.  A slide switch selects DC or AC coupling or it can be shorted to ground.

## Initial setup

The first thing to do upon power up, is to balance out any amplifier offset.  Here's the procedure:  At the upper right-hand side of the scope, you'll find the TRIGGER and SWEEP controls.  Set the TRIGGER control to INTERNAL and the SWEEP control to FREE RUN.  This will produce a horizontal line on the face of the CRT.  It sweeps from left to right and blanks out on return.  It doesn't matter what speed the sweep is running. just so it is moving to produce a horizontal line. It should look like Fig 4-11.

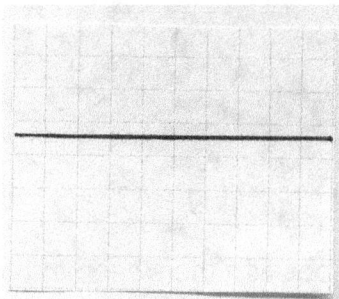

Fig 4-11

Now, back to the vertical amplifier.  Set the SENSITIVITY control to its <u>most</u> sensitive setting, and note the position of the sweep line on the graticle.  Now, set the SENSITIVITY control to its <u>least</u> sensitive position.  If there is a change in the position of the trace, use the BALANCE control to return it to its <u>original</u> position. The vertical amplifier is now balanced and it can be relied on for accurate readings.  You're now ready to use the scope as a test instrument.

## Voltage measurements

The SENSITIVITY control is a multi-position switch, calibrated in volts per centimeter.  Since the CRT gratical is scribed in 1 centimeter gradations, voltage can be read directly on the CRT screen.  Suppose you want to measure the output of a 5VDC power supply.  Switch the INPUT to DC, and the SENSITIVITY control to 1V/cm.

Set the SWEEP SELECT control to any position that will yield a steady horizontal line on the screen.  Use the POSITION control to place the sweep on any gratical line that is below center. (Fig 4-12)

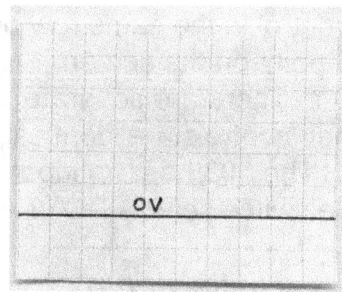

Fig 4-12

Probe the power supply output and note the position of the sweep. (Fig 4-12)  If the power supply is doing what it should, the sweep will move UP exactly 5 cm on the gratical.

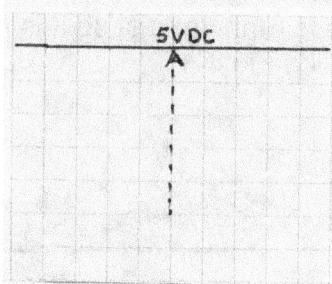

Fig 4-13

For AC measurements, position the sweep to the CENTER of the gratical and switch the INPUT to AC.  If the voltage to be measured is unknown, switch the SENSITIVITY control to its HIGHEST position.  Probe the voltage to be measured and note the peak-to-peak deflection on the gratical.  Adjust the SENSITIVITY control for an amplitude that is near to full scale on the gratical. (Fig 4-13)  Multiply the number of graticle lines spanned by the waveform by the position of the SENSITIVITY control.

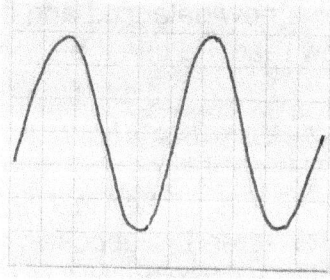

Fig 4-14

Let's, assume that the gratical deflection of the AC voltage is 2.5 cm above and below the center line, and that the SENSITIVITY control is set to 100mV per cm. What is the measured voltage?

Answer:  0.1X2.5=0.25V peak,  0.1X5cm=0.5V P/P.

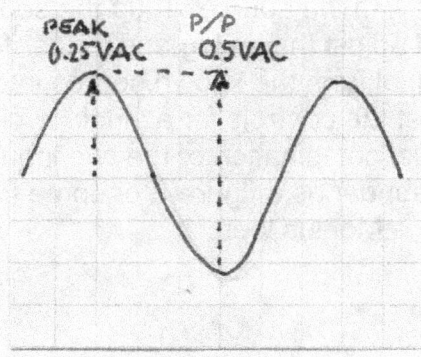

Fig 4-15

## *Frequency measurements*

The 503 and most quality oscilloscopes provide for a calibrated sweep time, (the time it takes for a single sweep of the electron beam to traverse the gratical). This gives us a way to measure the frequency of an input signal. Just to the right of the CRT is a control to set the SWEEP TIME per cm.

Suppose you are displaying the waveform of an unknown frequency. Adjust the SWEEP TIME control to display one full cycle of the waveform. Note the sweep time selected by the control. Multiply the SWEEP TIME number by the number of cm covered by one cycle of the waveform. This will give you the time it takes for one cycle of the displayed frequency.

Let's take an example. Suppose one full cycle of frequency covers 8 cm on the gratical and the SWEEP TIME selected is  10 msec per cm. (Fig 4-15) What is the frequency of the displayed signal?

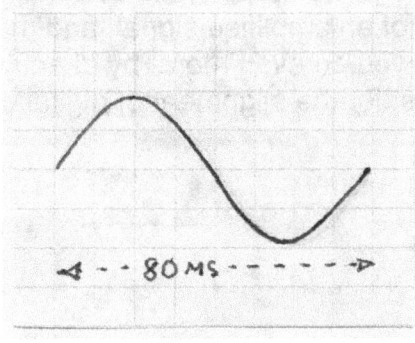

Fig 4-16

Answer: 0.001X8 = .008 sec. per cycle.  1/.008 = 125 Hz.

## *Frequency matching  (Zero-beating)*

Suppose you want to tune an oscillator to a specified frequency.  If you have an accurate frequency source to compare it with, you can use the oscilloscope to "zero beat" your oscillator with the source.  Plug the source into either the horizontal or vertical input adjust the timing and trigger for a single cycle of AC display.  Now, insert a signal from the known source into the remaining input of the oscilloscope and adjust the controls for a single cycle of source frequency.  What you will get is a visual combination of the two inputs called a Lissajous figure. (Fig 4-16). It may appear as a figure-8 or some combination of loops but it will be a closed figure like a Mobius loop.

LISSAJOUS FIGURE

Fig 4-17

Tune your oscillator until the display becomes as close to a figure "O" as possible.  The oscillator will now be at the same frequency as the known source.

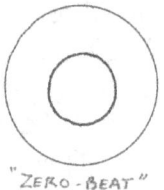

"ZERO-BEAT"

Fig 4-18

By now, you should be getting an idea of the power of an oscilloscope.  With it, you can measure such tings as the rise time of a square wave, the ripple voltage at the output of a power supply, the saturation and cutoff points of a transistor, the distortion of an amplified signal, and much much more.  A dual trace oscilloscope will let you do even more, by directly comparing the timing of two events--a must for the alignment of digital circuits. (Fig 4-19)

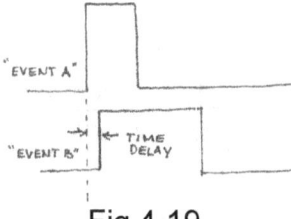

"EVENT A"

"EVENT B"

TIME DELAY

Fig 4-19

## *Digital oscilloscopes*

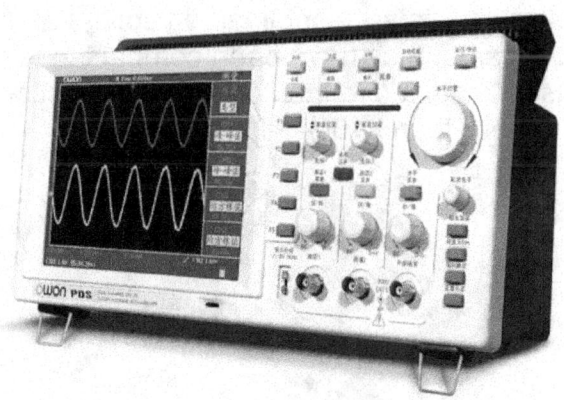

Fig 4-20

The digital age has opened up new possibilities for oscilloscope designers.. For one thing, since microchips are cheap and easy to assemble into complex devices, they have made it possible to acquire advanced features on a budget.

By their very nature, digital oscilloscopes are able to write images into memory and store them for future reference. It is also easy and relatively inexpensive to include one or more additional traces, making these instruments indispensable for analysis of clock and countdown circuits. You can also choose to display each trace in a different color. Here's a schematic for a typical digital oscilloscope

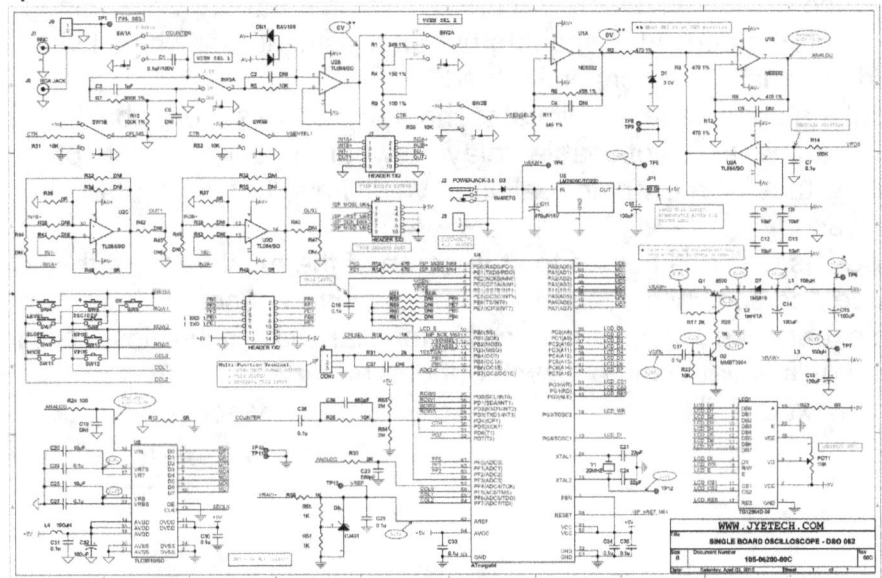

Fig 4-21

Analog inputs are buffered and converted to digital format. They are then processed by a CPU and displayed in graphic form on an LED or similar display. Because the necessary circuits are contained on solid-state microchips no bigger than a postage stamp, they have been compacted into a pocket size oscilloscope that was unimagined in the days of CRT O-scopes. This one is available in kit form for less than $100.

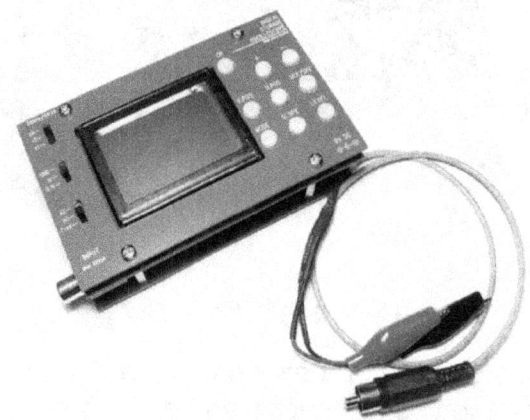

Fig 4-22

there is another class of O-scope known as the PC oscilloscope. It's a "black box", containing input buffering and D/A circuits. It comes with a disk containing a program that runs on your desktop PC or laptop. They are available pre-wired and in kit form for about $100 for a single trace 'scope or $200 for a dual trace 'scope. I use a single trace PC version made by Velleman. I find it meets my current needs well enough, but after using it for awhile, I wish I'd purchased the dual-trace version instead.

For all the advantages of digital technology, there's a drawback common to all digital instruments- a noticeable delay between an event and its presentation. if you must witness the event in real time, you'll have to revert to an old-fashion CRT oscilloscope.

I hope this Unit has helped to strip some of the mystery from the oscilloscope and convinced you to try one. A written description can't replace hands-on experience. Once you become familiar with an oscilloscope, I guarantee you'll never want to be without one again.

In the next unit, I'll introduce test oscillators and describe their use.

## *Terms to remember*

| | |
|---|---|
| **CRT** | Cathode-ray tube |
| **Free running** | Unsynchronized sweep of the electron beam |
| **Gratical** | Calibrated plastic "window" etched with reference lines. |
| **Sawtooth** | Linearly increasing voltage waveform with a sharp cutoff |
| **Schmitt trigger** | Circuit to prevent false triggering by noise. |
| **sweep** | Travel of the electron beam across the CRT screen |
| **Trigger** | Ability to start a sweep at a specific point on a waveform |

## *Exercises*

04-1     The sawtooth generator is set to sweep the electron beam at the rate of 1ms/cm. across a 10cm gratical.  What frequency of AC will display exactly one cycle on the CRT screen?

04-2     The vertical amplifier is set to display 0.1v/cm.  an AC waveform reaches 3.25cm p/p on the gratical.  What is its peak voltage?

# Unit-05, Audio Frequency Signal Generators

Fig 5-1

A signal generator is used to inject a signal into a circuit to check it's operation and to make adjustments to it. Next to the oscilloscope, a signal generator is the most useful tool in your arsenal of test equipment Actually, you will probably want several signal generators to cover as much of the audio and radio frequency spectrum as you'll be working with. Let's begin with the AF, Audio Frequency, portion of the spectrum.

## *Sine-Square Wave Generator*
An indispensable tool for checking audio amplifier performance is a sine-square wave generator. A good example is the *Heathkit* IG-18.

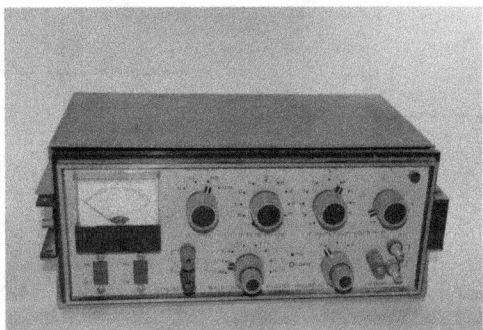

Fig 5-2

This one is many years old and has rendered faithful service in the design and repair of audio circuits for more years than I care to remember. Here's the schematic

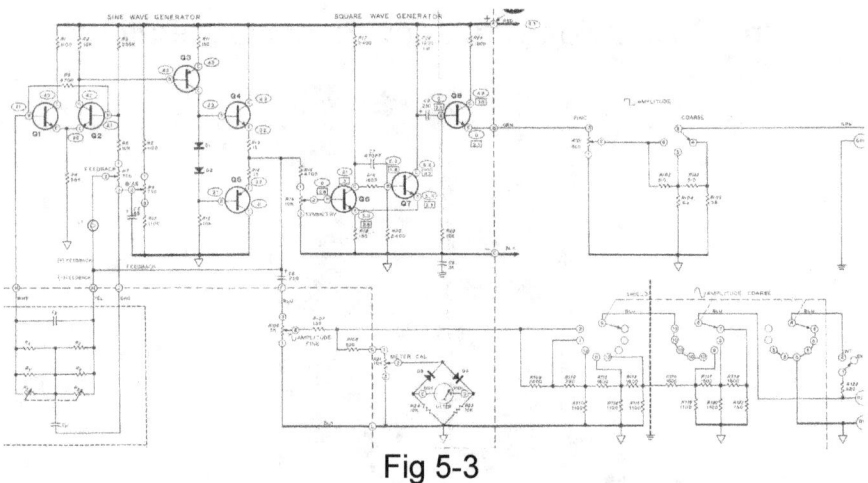

Fig 5-3

At the upper left is a very stable Wien-bridge oscillator, covering the DC to 100KHz range. Before I continue, let me take a side excursion into history. As I explained in Part-1, *Analog Circuits*, of this series, any amplifier circuit can be made to oscillate. Getting it to remain stable is another matter altogether. The goal of a test oscillator is to produce a pure waveform of constant frequency and amplitude. It's a design challenge that has plagued engineers from the very beginning.

The solution for a constant amplitude generator was found by the Prussian experimental physicist, Max Wein in 1891. He developed an oscillator incorporating both positive and negative feedback in an RC bridge circuit. Here's a modern transistor version of Dr. Wien's oscillator.

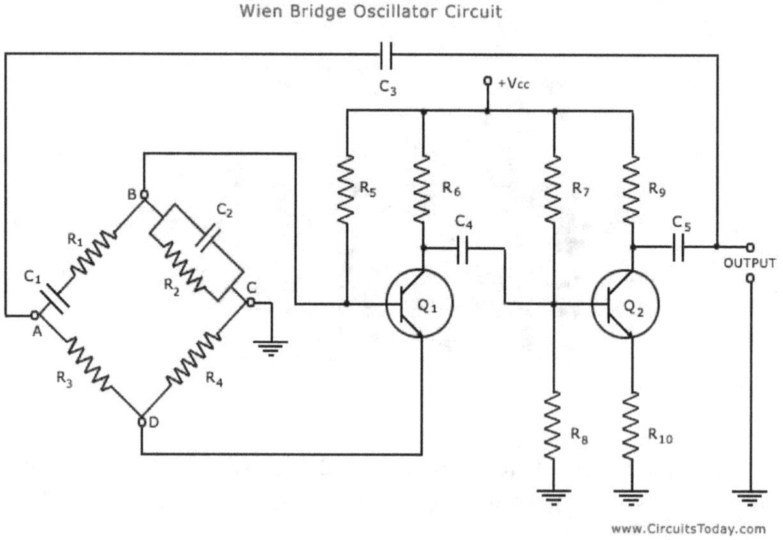

Wien Bridge Oscillator Circuit

www.CircuitsToday.com

Fig 5-4

Although better than its predecessors, the oscillator still had a problem with amplitude stability when confronted with a changing load impedance. In 1939, William Hewlett came up with a simple and elegant solution to the loading problem. What Hewlett did, was to replace one of the bridge resistors with a small incandescent lamp bulb (R-3).

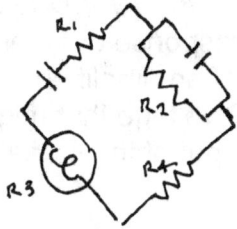

Fig 5-5

When the circuit oscillates, current is produced in the negative feedback leg of the bridge, heating the lamp. When the lamp filament gets hot, its resistance changes in such a way as to maintain a constant current. This limits the amount of feedback to keep the output at a constant level, making the oscillator very stable. with regard to amplitude.

Hewlett teamed up with Dave Packard to create the first commercial oscillators (Fig 5-6) to use the improved Wien bridge principle. If my memory serves correctly, *Walt Disney* studios purchased the first six oscillators to tune special, multi-channel theater sound installations for their 1940 film, *Fantasia*.

The 1939 Hewlett-Packard audio generators set the world standard for quality.

Fig 5-6

HP audio oscillators can sometimes be found on the internet  If you can get your hands on one, guard it carefully.  It's worth its weight in gold.

An audio signal generator, used in conjunction with an oscilloscope, can tell you a great deal about how an amplifier circuit is working.  Let's take some examples.

### Frequency response test

To measure the frequency response of an amplifier, capacitively connect the signal generator to the input of an amplifier stage.  The capacitor will isolate the signal generator from any DC voltage that might be present.  Connect the AC input of an oscilloscope to the output of the stage.

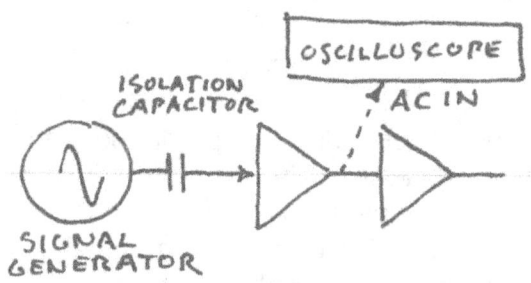

Fig 5-7

Adjust the output of the signal generator for a sine-wave that does not overdrive the amplifier and observe the output on the oscilloscope.  Adjust the image on the scope to cover several cm of gratical.  Now, without changing any of the settings, 'sweep" the generator frequency through the entire range the amplifier is designed to handle, and observe the amplitude of the waveform on the oscilloscope.  The signal should remain constant over a substantial part of the frequency range, dropping in voltage at the very low and very high ends of the "sweep"

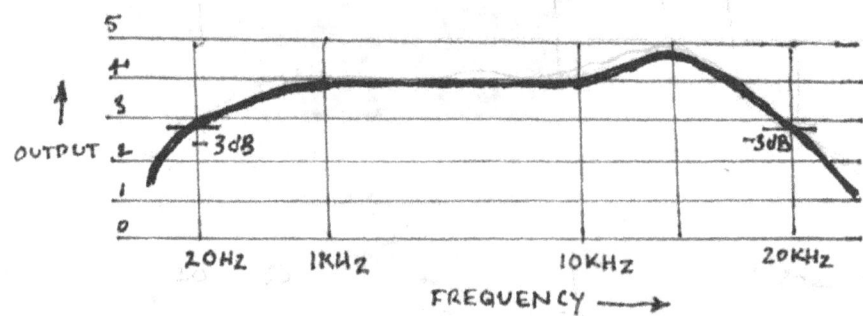

Fig 5-8

In the example in Fig 5-7, of output amplitude, plotted from about 10 Hz to 30KHz, it can be seen that the amplifier stage is only "flat" from 1 KHz to 10KHz. It has a sharp peak at about 12 KHz and it drops by 3dB at 20 Hz and at 20 KHz, not bad for a communications amplifier but not particularly good for music reproduction. This sort of response is difficult to judge by ear but with the right tools, it can be measured precisely.

Audio generators that could be made to electrically sweep a broad portion of the spectrum used to depend on tuning capacitors and motor drives. Today, you can download software that will use your PC sound card to generate swept audio signals. The waveforms are digitally generated and converted to audio with a D/A converter.

A word of caution. The sound card built into your PC tower or desktop is not very precise or very stable in performance. If you want to use a computer to generate precise audio signals, you should install a professional sound card made for MIDI use. The clock and AD/DA converter will ensure accurate results. These cards normally sell in the 200 to 300 dollar range, but you can often find a used card for about 25 to 30 dollars.

## Distortion test

Now, let's take a look at the amplifier's linearity. Connect the signal generator and oscilloscope as you did to measure frequency response. Set the generator frequency somewhere near the middle of the amplifier's operating range. Begin with the signal generator at its lowest output level, then increase it gradually, until the signal begins to distort. The waveform should remain :pure" as the amplitude is increased, and should reach cutoff and saturation simultaneously.

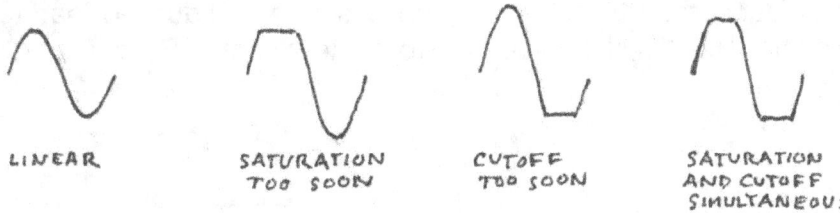

Fig 5-9

If either saturation or cutoff is reached too soon, the amplifier will distort at high volume levels. You'll need to check of each amplifier stage until you find the one that is operating outside its linear range.

Adjust the bias voltage on one or more transistors (vacuum tubes, if you're working with an older amplifier) for linear output . Intermodulation distortion is a common ailment in amplifiers of all sorts. An amplifier stage can exhibit a "fuzziness" or "ringing" when certain frequencies are input. This is usually caused by very high frequency oscillations within the amplifier. Oscillations may be present but above the hearing range, or they can be initiated by sharp transients, such as symbol crashes in music, or by sudden volume spikes. This is where the square wave generator becomes useful.

Connect the square wave output of the signal generator to the input of the amplifier being tested, and observe the output signal as you did the sine wave. Be sure to keep the input level low enough to operate the amplifier in its linear range. The square wave will contain harmonics of the fundamental frequency of the oscillator. An amplifier with good transient response will produce an output waveform that is close to the waveform of the input. In any event, if there is a tendency for some harmonic to induce an oscillation in the circuit, this test will reveal it.

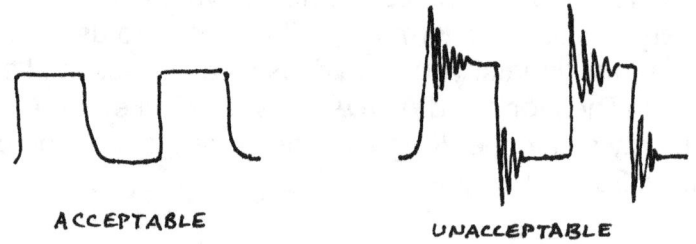

ACCEPTABLE          UNACCEPTABLE

Fig 5-10

### RF carrier modulation

If you use a microphone or other audio input with a transmitter it is essential that you do not over-modulate the carrier. To do so can cause the RF signal to to be radiated outside the intended band. A useful tool to check transmitter modulation is a two-tone audio generator. Connect the two-tone oscillator to the microphone input of the transmitter and an O-scope and dummy load to the output. In a later Unit, I'll show you how to make a simple RF coupling transformer.

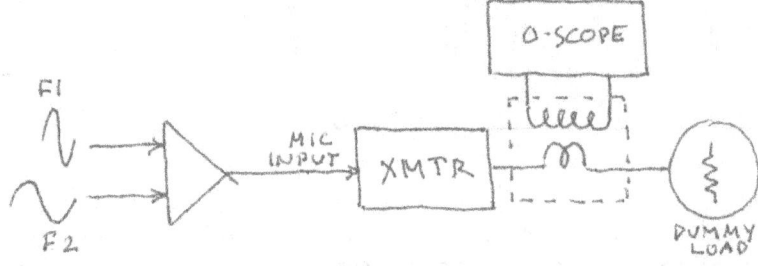

Fig 5-11

By applying two audio tones of different frequencies to the microphone input and observing the transmitter output on an oscilloscope, You can see if the signal is being properly modulated. A two-tone modulated RF carrier looks like this.

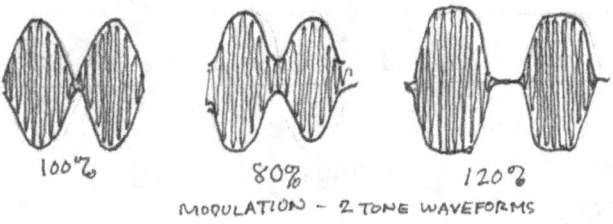

Fog 5-12

A two-tone audio generator is not likely to be high on your priority list. Not to worry. There are freeware programs available that will turn your PC and sound card into a two-tone generator. Here's a picture of the PC screen for one of these programs.

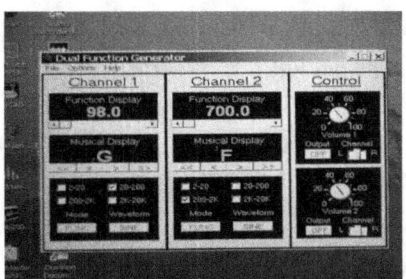

Fig 5-13

Once downloaded into your computer, the program behaves like a piece of hardware with dials and pushbuttons. You set the frequency and output level of the audio tones with the PC mouse. The audio tones can be selected separately or combined into a single output. the program shown in Fig 5-13 offers Sine, square and triangular waveforms and also "white" and "pink" noise.

Caution - sound-board generated waveforms are digitally synthesized and therefore quite accurate as far as frequency goes but unless they are "aliased" to smooth them out, they contain millions of sharply defined steps yielding harmonics that have the potential to interact with the circuits you are measuring.

Fig 5-12

In this unit, I described audio signal generators and their use in troubleshooting audio circuits.  In the next unit, I'll discuss RF signal generators.

## *Terms to remember*

**Wien bridge oscillator**          Oscillator circuit using both positive and
negative feedback

**Frequency response**          Voltage or current gain of an amplifier stage as
a function of frequency

**Linearity**          Distortion of an amplifier stage output with
regard to its input

**Intermodulation distortion**          Generation of spurious signals when  a signal
is "beat" with harmonics or internal oscillations

**synthesized waveform**          Waveform generated by digital means rather than by
an oscillator

**Two-tone oscillator**          Oscillators generating audio tones of different
frequencies, combined into a single output.

# Unit-06, Radio Frequency Signal Generators

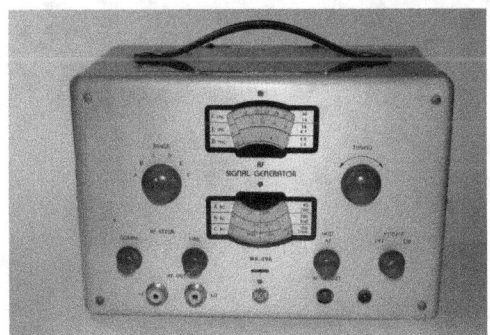

Fig 6-1

The alignment of the RF and IF stages of a receiver is greatly facilitated with the use of an RF signal generator.  In the vacuum tube era, it was not an easy task to obtain a signal at a specific frequency for long.  Vacuum tube oscillators tended to drift with temperature.  Even crystals, stable as they are, had to be maintained at a constant temperature to guarantee their physical dimensions, and therefore their resonant frequency..  Few repair technicians could afford a really good RF signal generator.  Fortunately, there were a few kits on the market that were capable of providing a "ballpark" estimate of frequency and were good enough for the alignment of AM and early FM radio receivers.  Typical of the genre was the Heathkit RF-1 (Fig 6-2).

Fig 6-2

It covered six bands: 100-320KHz, 310-1100KHz, 1-3.2MHz, 3.1-11MHz, 10-32MHz and 32-110MHz. and contained an internal 400Hz oscillator to modulate the RF signal.  It was a simple VFO using two vacuum tubes   Here's the complete schematic.

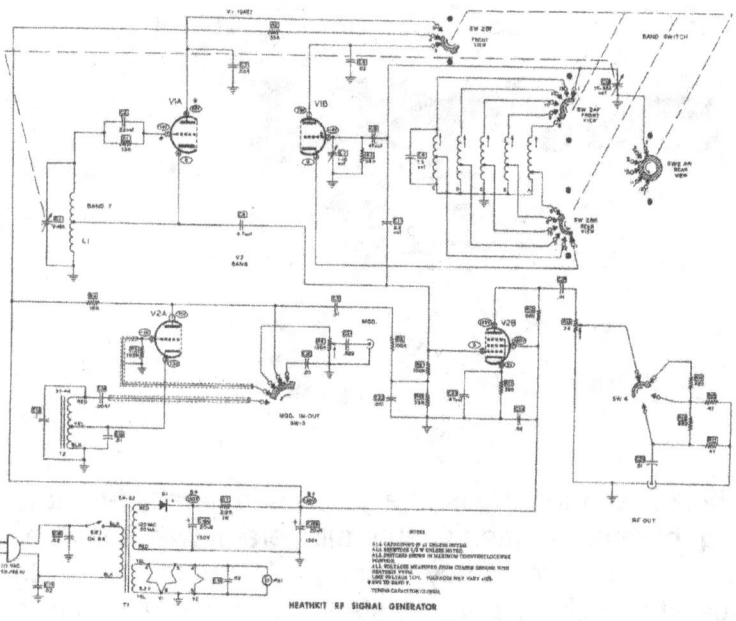

Fig 6-3

The vacuum tube circuit is a Hartley oscillator. L1 through L5, in parallel with tuning capacitor C12, are selected by switch S2 to determine five of the 6 operating bands. The 32-110MHz band is wired, separately, to V1A in order to keep the connections as short as possible and to eliminate stray capacitances. Switch S3 selects the carrier modulation source (OFF,400Hz, Ext). V2A is The 400 Hz oscillator. The mixer stage is V2B  To make it easier to visualize, here's a simplified schematic of the RF oscillator portion of the signal generator

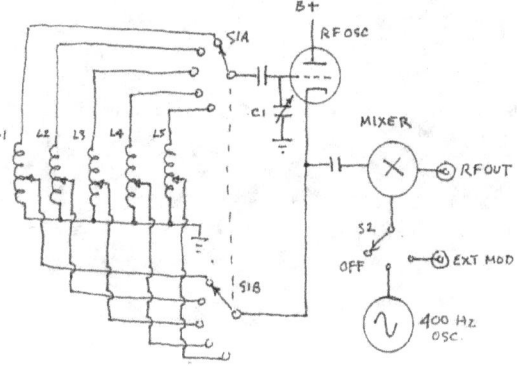

FIG 6-4

The vacuum tube signal generator served a purpose 50 years ago, and there are a goodly number of these relics available on the used equipment market. I own a nearly identical version of this generator made by RCA. I still use it.

Newer RF signal generators are solid state and produce synthesized waveforms, making them stable and accurate. They can also be programmed to sweep over a selected frequency range and even be FM modulated. Fig 6-5 is an example of a synthesized RF generator. Notice the lack of a tuning dial. Everything is selected by pushbuttons and the output frequency is displayed by an alpha-numeric readout.

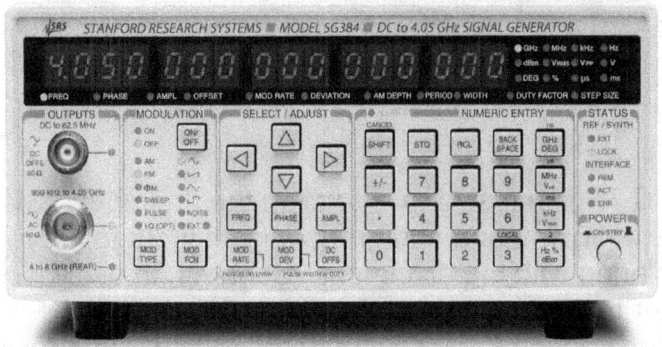

Fig 6-5

The architecture for a synthesized generator is quite different from the standard vacuum tube or transistorized generator. For one thing, there is no oscillator in the traditional sense of the word. Algorithms stored in memory, are used to synthesize output waveforms from a table of values. The selected values appear as a stepped series of voltages that are filtered and passed to a digital to analog converter. The analog waveform is further filtered and presented as a shaped waveform of a desired frequency. Here's a simplified block diagram of the process.

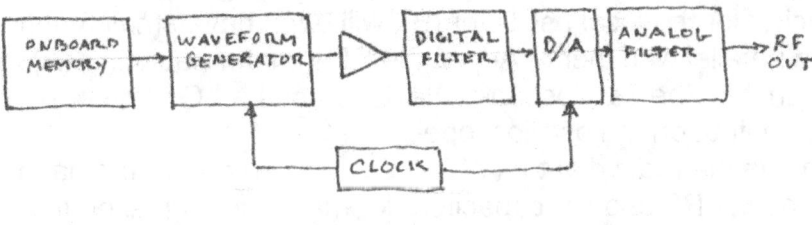

Fig 6-6

So, now we have an RF signal generator at our disposal, whether it be a vacuum tube or solid state model. How do you use it to align a radio? Let's take an AM broadcast receiver as an example.

## AM Receiver Alignment

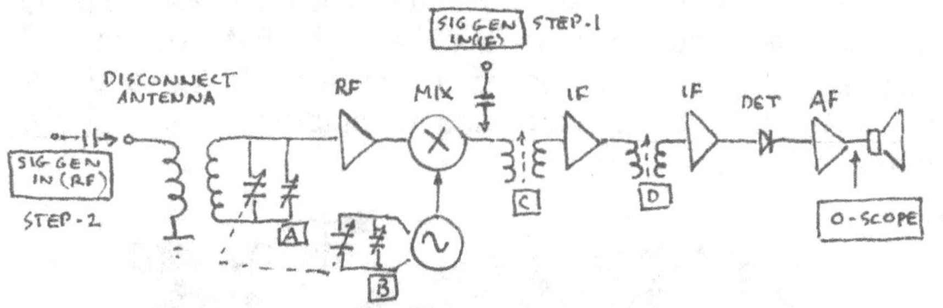

Fig 6-7

Refer to Fig 6-7. The first step is to align the IF stages. Start by disconnecting the radio antenna and grounding its input to prevent off-the-air signals from interfering with the alignment. Connect an oscilloscope or a sensitive AC meter to the to the radios' output. Connect an RF signal generator, through an isolation capacitor, to the input of the first IF stage coupling transformer. Tune the signal generator it to the radio's IF frequency (455 KHz). Modulate the generator with an audio tone. Begin with the lowest amount of IF signal that will give a reading on the oscilloscope. Adjust the tuning slugs or trimmer capacitors on IF transformers (C) and (D) for maximum signal on the oscilloscope. Be sure not to drive the IF stages into saturation. The IF stages will now be properly tuned.

The next step is to adjust the RF stages and Local Oscillator. Tune the radio dial to a point near the high end of the band (e.g.1550 KHz) Connect the signal generator to the antenna input through an isolation capacitor. Tune it to 1550 KHz and modulate it with an audio tone. Start with just enough signal to get a reading on the oscilloscope. The main RF tuning capacitor (A), and the tuning capacitor for the local oscillator (B), will each have small trimmer capacitors in parallel with them. Adjust the RF trimmer capacitor (A) for maximum signal on the oscilloscope. Next, adjust the LO trimmer capacitor (B) for maximum output on the oscilloscope.

Next, move the radio dial to a point near the low end of the band (e.g.550 KHz). Re-tune the RF and LO capacitors for maximum signal on the oscilloscope. Depending on the tracking accuracy of the RF and LO tuning capacitors you may have to compromise between the two band ends to obtain a satisfactory overall tuning.

If you don't have access to an oscilloscope or other sensitive measuring instrument, you can do the adjustments by ear, listening to the output of the modulated signal from the radios' loudspeaker. The adjustments won't be as accurate, but they should be fairly close.

## FM Receiver Alignment

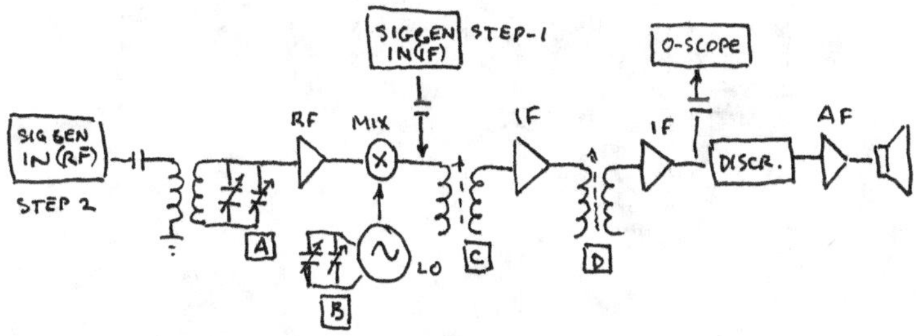

Fig 6-8

Except for the operating frequency, FM receiver RF, IF and LO stages can be tuned exactly the same way as you would for AM receivers. The only difference is that, unless the signal generator is capable of FM modulation, the carrier won't be detected and passed to he AF stages. Therefore, you'll need to connect the O-scope to the output of the last IF stage in order to read the signal.

Refer to Fig 6-8. Begin with step-1 as you did for the AM receiver, For a broadcast FM receiver, connect the signal generator to the input of the 1st IF transformer (C) set the RF signal generator for an IF frequency of 10.7MHz. Set the oscillator level just high enough to break the squelch of the FM receiver. Adjust the IF transformers for maximum output. The IF stages should now be set for a center frequency of 10.7MHz. This will not tell you anything about their bandwidth but the adjustment should be adequate for normal reception.

Next, tune the RF and LO stages the same way you did for the AM receiver. Start with the receiver dial and signal generator for a frequency near the high end of the FM band (e.g. 108MHz), and then the low end (88 MHz).

If your signal generator can be FM modulated, you can adjust the discriminatormcircuit for a symmetrical waveform centered on the IF frequency of 10.7MHz. If not, it's best not to try a discriminator adjustment unless there is significant distortion of the audio signal.

## Swept Signal Generators

The ability to automatically sweep a constant amplitude RF signal over a broad frequency range enables one to view the overall response of an amplifier stage from its lowest to its highest operating frequency without having to make detailed plots at a great number of individual frequencies.

One method of creating a swept RF signal is to beat a master oscillator output with a voltage controlled local oscillator and combine the two signals in a mixer to provide a broad-band swept output. It diagrams something like this.

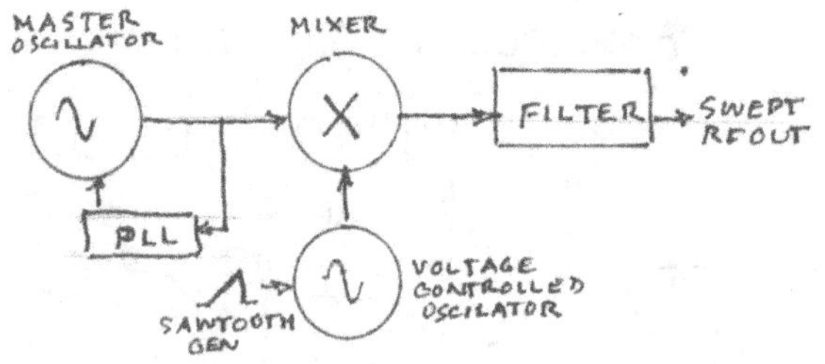

Fig 6-9

This approach uses a PLL stabilized master oscillator to generate a reference frequency.  A voltage controlled oscillator, operating very much like the local oscillator in an AM  radio receiver, but controlled by a sawtooth tuning voltage, is mixed with the MO output to produce two signals, equal to the sum and difference of the two oscillator frequencies.  The output is filtered to eliminate the unwanted signal, resulting a swept RF output of the desired frequency range.  The bandwidth of the output signal is determined by the amount of sawtooth voltage applied to the LO and the sweep rate is determined by the length of the sawtooth.  Since digital signal generators synthesize their waveforms, it is easy to make them into a swept RF generators.

So, why would you want a swept oscillator?  Because, when used in conjunction with an oscilloscope, it can give you a visual image of how a piece of equipment behaves within a limited frequency range.  Let's take a couple of examples.  Here's what the response of a properly tuned IF stage looks like on an oscilloscope screen when the signal generator frequency is swept across the IF passband.

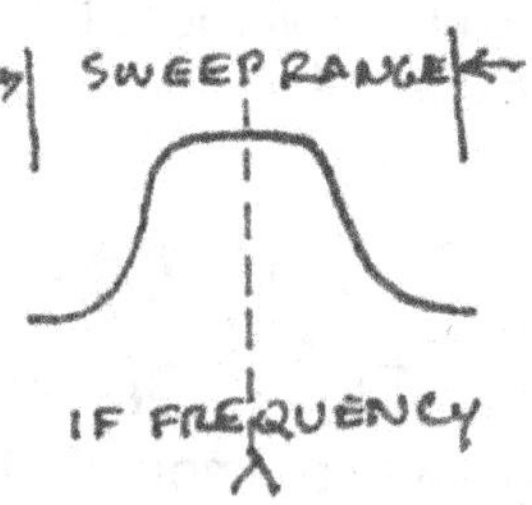

FIG 6-8

As you can see, the ability to see the response of the IF stage to a swept signal makes it a cinch to tune the stage for optimum amplitude and bandpass.  Now, let's see how a swept oscillator and oscilloscope can show the response of an FM discriminator

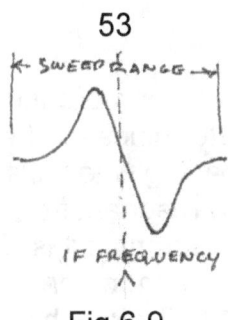

Fig 6-9

Once again, it becomes obvious that a real-time image of the discriminator response makes the job of tuning for perfect symmetry an easy task.

Besides the standard bench-type signal generators, there are a host of specialized RF generators. Depending on your needs, these might include one or more of the following examples.

## *The Dip Meter*

Fig 6-10

The Dip Meter, (in the age of vacuum tubes it was called the Grid Dip Meter), is useful to "dry-tune" a reactive circuit or to find the resonant frequency of a reactive circuit. It is a low power hand-held oscillator with an externalized tuning coil. Here's a typical circuit.

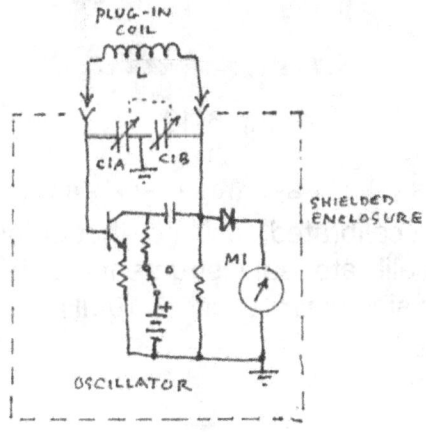

Fig 6-11

Plug-in coils cover a half-dozen or more tuning bands and a calibrated dial, attached to a tuning capacitor, determines the frequency of oscillation. The oscillator is sensitive to current loading when it is brought near a resonant circuit. When the oscillator is tuned to the resonant frequency of the circuit, base current in the oscillator transistor (or grid current, if it is a vacuum tube oscillator) dips to a minimum as registered on a micro-ammeter

Refer to the circuit in Fig 6-11. To use the Dip Meter, plug in a coil (L) for the frequency range you are interested in and switch ON battery voltage to the oscillator. Hold the external coil near the circuit you are testing and tune capacitor (C1) until the current reading on the meter (M) dips to a minimum. The dial reading on the dip meter will indicate resonant frequency of the circuit being tested. It's important not to couple the Dip Meter too closely to the circuit as this will cause too much loading on the oscillator and will tend to detune it to give a false frequency reading. Hold the coil just close enough to the circuit to get a slight dip at resonance.

Needless to say, the Dip Meter is not a precision instrument. It is used only to get a "ballpark" idea of resonance. Some Dip Meters have internal AF modulators and can be used to adjust radio AF and IF stages much the same as an RF signal generator. You only need to hold the Dip Meter near a radio to inject signals into the RF or IF stages being adjusted. It's not as accurate as a precision RF signal generator but it's a lot cheaper. Some Dip Meters even have an output co connect a frequency meter so that they can be read with more precision.

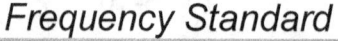

## *Frequency Standard*

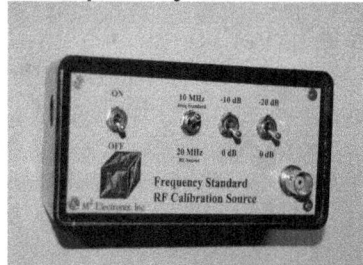

Fig 6-12

An expensive but often indispensable tool for the technician is an RF frequency standard. It is a calibrated oscillator that outputs a signal at a precise frequency and is used to calibrate test instruments such as frequency counters. This one, by $M^3$, outputs a sine wave at either 10MHz or 20MHz. The basic circuit looks something like this:

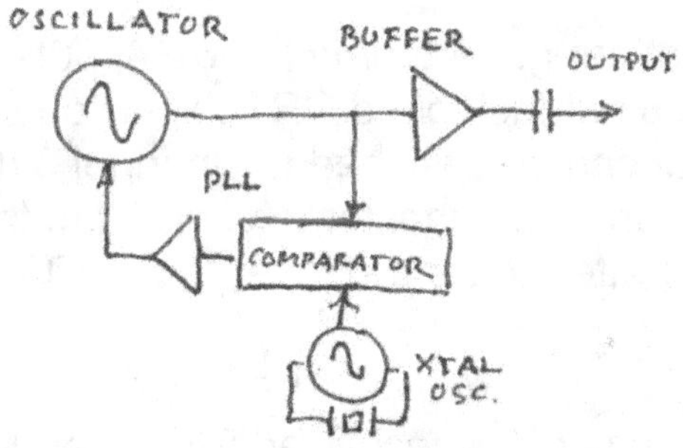

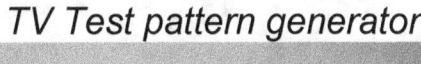

OSCILLATOR    BUFFER    OUTPUT
PLL
COMPARATOR
XTAL OSC.

Fig 6-13

The output of a master oscillator is sampled and fed to a circuit that compares it with the output of a very stable crystal reference source. Any difference between the two signals produces a voltage that is used to control the frequency of the master oscillator to lock it in phase with that of the crystal. This is called a phase-lock loop. It is accurate to ±1Hz.

Besides calibrating test instruments, the frequency standard can be used to input a known frequency to a communications receiver to calibrate it.

Barring access to a frequency standard, it is customary to calibrate an amateur transmitter to National Bureau of Standards radio station WWV in Colorado, or WWVH, Kaua'i, in the Hawiian Islands. These broadcasts are continuous and include a number of useful signals for calibration purposes. Broadcast frequencies are 2.5, 5, 10, 15, and 20 MHz. The method is to tune the transmitter to zero-beat with WWVs carrier, Note the carrier offset from the dial on the transmitter, and re-tune the transmitter's Master Oscillator to bring the dial indication and WWV's carrier into alignment. It is difficult to do accurately so the availability of a precision standard that can substitute for WWV makes the job a lot easier.

However you use a frequency standard, treat it with great respect. Keep it in a temperature stable, padded container when not in use. And don't drop it!

## TV Test pattern generator

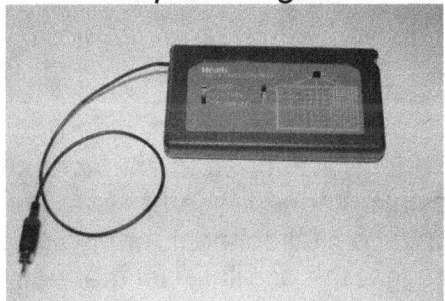

Fig 6-14

The *Heathkit* test pattern generator in Fig 6-14, puts out a several black and white test patterns to align old-fashioned CRT television sets. It generates a modulated RF carrier that can be switched to Channel 3 (61.25MHz) or Channel 4 (67.25MHz). Output is an "F" connector to connect to the TV set's antenna input. The selected test patterns are displayed on the TV screen

## *Antenna Analyzer*

Here's another example of a specialized RF generator. It's an Antenna Analyzer, capable of determining the standing wave ratio (SWR) and impedance of an antenna within the HF to VHF range.

Fig 6-15

In its simplest form, the antenna analyzer is a variable frequency RF generator driving an impedance bridge

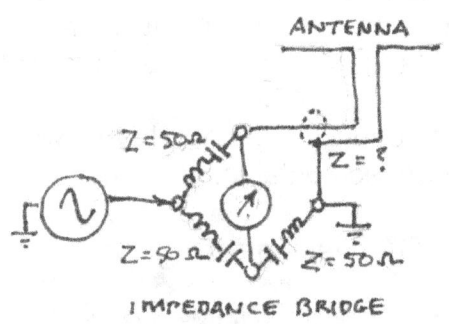

Fig 6-16

. Three sides of the bridge are tuned to an impedance of 50 ohms (the common standard for coaxial antenna feeders). One leg of the bridge, however, is missing. It terminates in a coaxial connector to which the antenna to be tested is attached. To use the bridge, the oscillator is manually swept through the band segment the antenna is presumed to cover. The bridge will balance (and therefore yield a minimum meter reading), when the antenna is at resonance and its impedance is 50 ohms.

Reflected power from the antenna is amplified, converted to a digital value, and fed to a CPU. The CPU calculates the impedance and reactance of the antenna and displays them directly. It makes the job of tuning an antenna child's play.

There are some caveats, however. For one thing, the length and quality of the transmission line must be taken into consideration. If the line is essentially lossless,. the readings can be considered accurate within acceptable limits. If there is significant loss in the transmission line, or mismatch in a balun transformer, the readings will be affected. Best to omit any transmission line or balun transformer and connect the antenna directly to the instrument.

I have found that tuning an antenna in a location where it will not be used can give a false reading. I find it useful to take the readings with the antenna is mounted in its final position, using as short a length of transmission line as necessary to reach the antenna from a convenient place to stand. You may have to put the antenna up and take it down several times before a final tuning is achieved.

The analyzer in Fig 6-15 is made by MFJ and, because it contains a CPU, it can also be used to adjust antenna tuners, matching networks and balun transformers, and it can measure the impedance and velocity factor of a transmission line. It is pricey but, as you can see, it is a rather versatile instrument--a must, if you experiment with home-brew transmitting antennas.

This completes the unit on RF signal generators. There are more types of specialized generators but the ones mentioned here are the ones you'll most likely need. In the next unit, I'll introduce frequency counters.

# Unit-07,-frequency counters

Fig 7-1

An absolute must for adjusting the master oscillator in a modern transceiver is a high resolution frequency counter. It can be a megabuck laboratory model with all the bells and whistles or it can be a simple kit you can assembly yourself for less than a hundred dollars. Remember, though, that the accuracy of the counter will determine the accuracy of your measurements. The counter must be calibrated against a known standard for it to be of any use. Lets begin with a block diagram of a typical frequency counter.

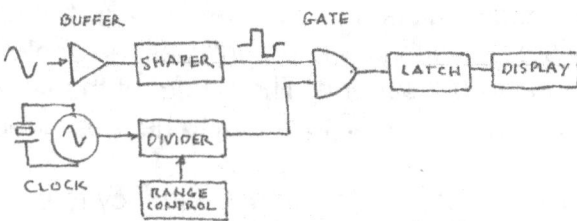

Fig 7-2

The unknown signal is amplified and then shaped into a sharp rise-time square wave to make it easier for the digital circuitry to respond to the waveform. This does not alter the frequency of the signal. It simply transforms the sine wave into a square wave of the same frequency. All of the better counters incorporate a Schmitt trigger to eliminate false readings from noise spikes. Here's the input amplifier circuit from a vintage *Heathkit* counter.

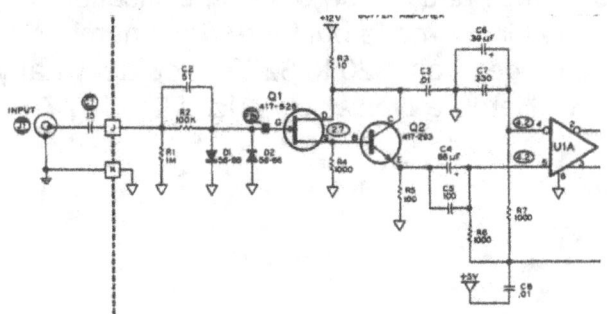

Fig 7-3

The input stage is a field effect transistor, providing a high impedance to prevent the counter from loading the circuit it is measuring. Diodes D1 and D2 limit high level signals by clipping them to prevent accidental damage to the FET.

A high frequency crystal controlled clock produces a pulse train to synchronize events within the counter. The clock pulse train is "counted down" and fed to one input of an AND gate. Here's the master oscillator and countdown circuit from the *Heathkit*.

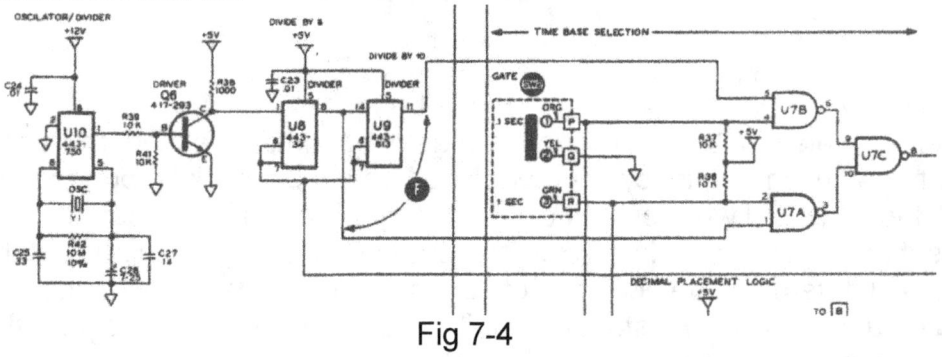

Fig 7-4

The crystal controlled clock is at the far left. Integrated circuits U8 and U9 are the frequency dividers. Switch SW2 determines the scale for gate timing, in this case either 1 second or 0.1 second. The output of the selected frequency divider is applied to one gate input. The signal to be measured is fed to the other gate input.

Now, let's suppose you expect the input frequency to be 850 Hz and you choose the 1 sec gate timing. That would result in a countdown of the clock such that the gate would be "open" for a total of 1 sec), allowing however many cycles of the unknown frequency to pass. It would then shut down until commanded to re-start. The pulses of unknown frequency are latched in a shift register and displayed on a digital readout.

Let's further suppose that the readout has a resolution of 8 places and a tolerance of ±1 part per million. If the actual frequency of the unknown signal is 850 HZ the display on the counter can be expected to show 850.00000. On the other hand, if the display were good to 9 places. it could show anything from 849.999 999 to 850.000001 and still be within specification

The *Heathkit* is no longer made but there are a number of frequency counter kits available in price ranges from $20 to $200  You get what you pay for in both accuracy and stability. Here's a typical low priced kit. (Fig 7-5)

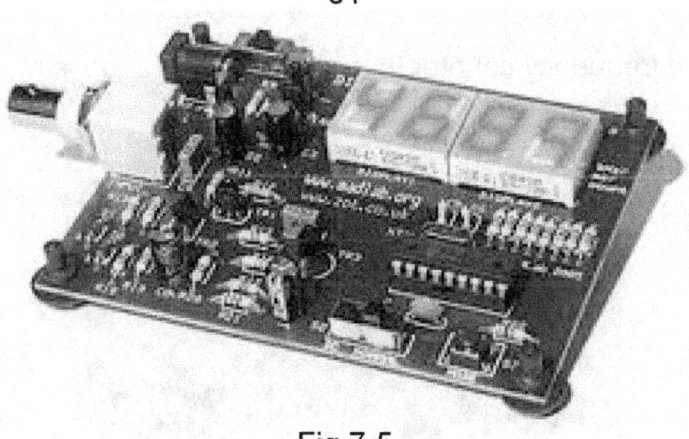

Fig 7-5

The kit is built designed around a TCXO self-contained oscillator package and a PIC microcontroller chip. It has two range settings, 0-30 MHz with a resolution of ±10 Hz, and 0-3000MHz with a resolution of ±1000 Hz. These aren't particularly good specifications but, hey, what do you expect for a sawbuck?

PIC microcontrollers are a wonder of the modern age. In a single monolithic package, you get what amounts to a personal computer the size of a nickel, and it can be programmed to do just about anything imaginable. Here's a block diagram of the PIC used in the frequency counter (Fig 7-5).

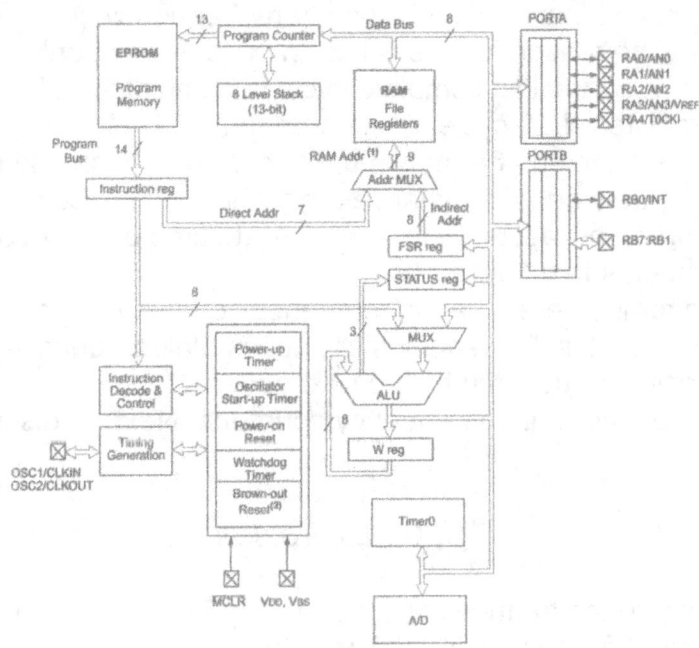

Fig 7-6

Assembling one of these kits is not for the faint-heated. You need a fine point temperature controlled soldering iron and a steady hand.

A very good frequency counter in the $100 price range is made by M³. company (Fig 7-7)

Fig 7-7

The counter is good to 1 part in 1,000,000,000 and comes fully assembled and calibrated or you can build it as a kit. Unless you are willing to opt for their 10/20 MHz calibration standard (described in Unit-6), I don't recommend you build one. You'll need to calibrate the clock with standard or the counter will be useless.

It should be noted that the resolution of any frequency counter is dependant on the gate time. The reason for this is that the counter can only count complete cycles of frequency, so the last complete cycle accumulated before the gate "closes" will be registered. A cycle that has begun but is not complete upon gate "closing" will not be counted. For maximum accuracy, you should use the longest available gate time. For instance a gate time of 1 second will allow frequency measurements accurate to 1 Hz. a gate time of 10 seconds will resolve measurements to 0.1 Hz.

It should be remembered that no frequency counter will provide measurements in "real time". Readings are accumulated during a gate cycle and are latched in place until updated by the next gate cycle.

This completes the Unit on frequency counters. Next, I'll discuss power supplies.

## Unit 07 Exercises

07-1    Refer to the frequency counter in Fig 7-5. What might the display show for a measured frequency of 1.5MHz if the counter were set to the 300MHz scale?

07-2    What would the same display show if the counter were set to the 30MHz scale?

## Unit 08, Power Supplies

Fig 8-1

If you're an experimenter or circuit designer, or if you simply want to bench test a piece of equipment, you'll need to power it with something. Often, just a battery will do but just as often, you'll need some kind of variable output bench supply. What kind of supply will, of course, depend on the circuit you are dealing with. Most modern electronic circuits run on 5, 9 or 12 volts DC. This is because they are built from off-the-shelf modules that are standardized to operate from common power sources.

If you are primarily interested in servicing computer circuits, a fixed supply of +5VDC and ±12VDC will cover about 90% of your needs. The cheapest solution is a standard ATX power module found internally in any PC tower. Before scrapping that old PC tower, you might want to remove the power supply module and convert it to a bench instrument by adding a power takeoff socket of some sort. Here's an example of an ATX power module.

Fig 8-2

It is a well regulated high current source of +5VDC and ±12VDC.

Here's the color code for the output leads and a schematic for how to connect them for bench use.

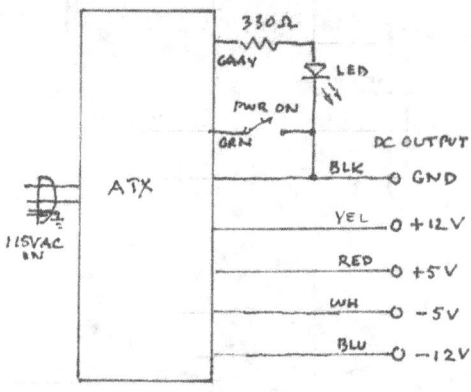

Fig 8-3

Yellow   +12VDC at about 12A
Blue      -12VDC at about 8oo mA
Red       +5VDC at about 30A
White     -5VDC at about 0.5A
Black     common ground

Wire a SPST switch between the green and black wires as a power ON switch and you can power an LED ON lamp in series with a 330Ω 1/4 watt resistor between the gray and black leads.  I should caution you that  ATX  power modules are switching mode power supplies.  Older models tended to put out quite a bit of RF "hash" --not necessarily bad for computer use, but not clean enough for some circuits.  Newer models have better filtering at both the AC input and DC outputs

If you need a variable voltage, A number of good power supply kits available at reasonable prices.  Here's one that offers 2 independent supplies (each variable from 0 to 30VDC at 3A) in a single package -- a very nice combination for the circuit developer.

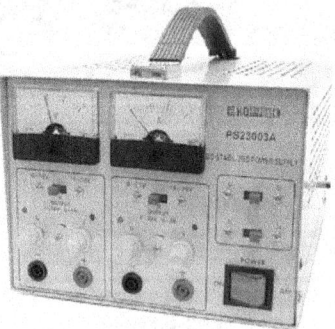

Fig 8-4

If you prefer a commercially built power supply, there are both transformer type, and switching mode supplies available in a large range of fixed and variable voltages. This transformer type power supply, made by *MFJ*, has been powering my radio shack for the past five years. It produces 30A of well regulated DC with good line and load regulation.

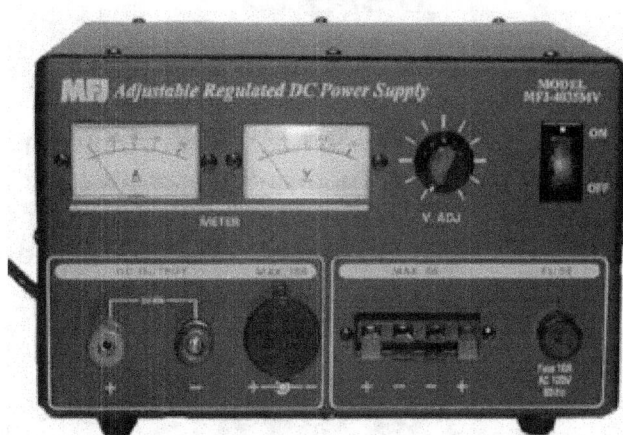

Fig 8-5

While I'm on the subject of power supplies, line regulation can be checked using a variable transformer. They are commercially available under the names *Powerstat* and *Variac*.

Fig 8-6

These transformers are not cheap but they are very useful. They are classified as single-winding autotransformers. Here's a typical schematic

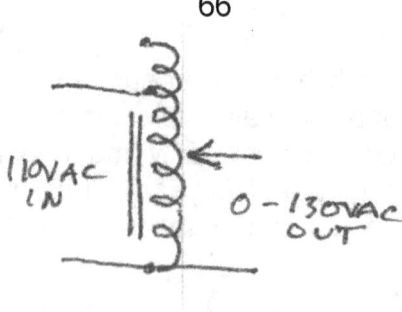

Fig 8-7

Simply connect the equipment under test to the output and vary the voltage over the specified range to see if the supply regulates properly.

Another important power supply specification to watch for is ripple and noise output. This is usually specified in mV and not easy to verify, especially if the power supply has a high voltage output. One way to verify a small noise signal on a large voltage is to compare the supply to a stable source of nearly equal value.

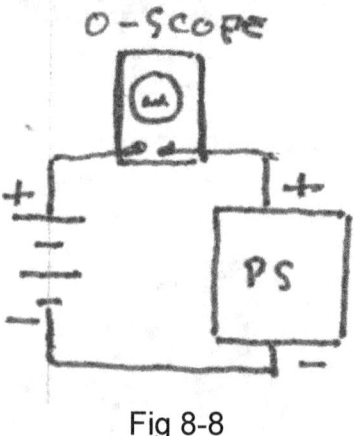

Fig 8-8

Suppose you want to check the output of a hundred volt power sjupply. If you measure the supply output with an o-scope, you can set it to measure AC and isolate it with a capacitor. This will read the AC ripple on the output, but it will not see small DC voltage changes. You can connect eight 12V batteries in series to make a stable 96V source. Now, you can "float" an oscilloscope between the two power sources. The difference between them will be 100-96 = 4 VDC. Connect the DC vertical input of the o-scope so that the positive lead is on the battery stack negative terminal and the negative lead is on the + output of the power supply. You can now read the stability of the power supply output on the oscilloscope screen. Any drift in the power supply output voltage or any ripple or noise present at the output can be observed directly

## Unit-9, Some additional items

### Substitution Boxes

If you're an experimenter, a lot of time can be saved by quickly substituting one value of a component for another to see what the effect on a circuit will be. Here's an example of a now-extinct Knight kit capacitance substitution box.

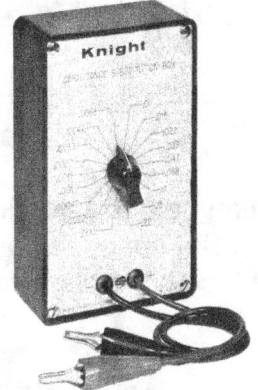

Fig 9-1

The box is a pretty simple affair -- standard value capacitors from .0001µf to 0.22µf are connected to a pair of test leads by a rotary switch. Here's the schematic.

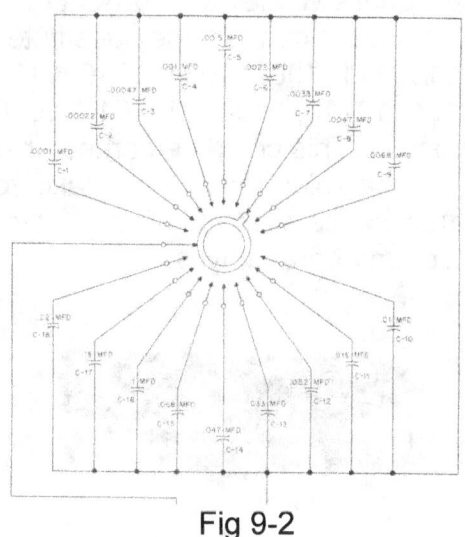

Fig 9-2

The Knight kit is no longer available but, if you have a well-stocked junk bin, you can easily construct your own in a couple of hours.

The same can be said for a resistance substitution box. Consider the possibilities offered by these examples.

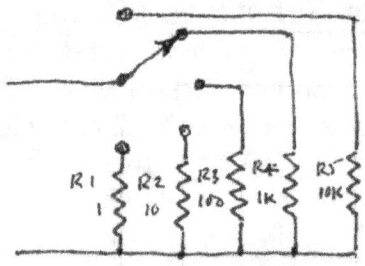

Fig 9-3

Like the capacitor substitution box, this version provides standard resistors arranged in decade progression.  But suppose you are designing a critical circuit that requires a finer cut of resistance?  Here's a version that will help fill in the gaps.

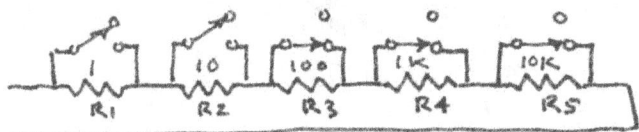

Fig 9- 4

This version uses individual switches to short-out resistors in a series string. With only five resistors in decade, it can provide substitute resistance values of 1, 10, 11, 100, 110, 101, 110, 111, 1000, 1001, 1010, 1100, 1100, 1111, 10000, 10001, 10010, 10100, 10101, 10110, 11000, 11001, 11010, 11011, 11100, 11101, 11110 and 11111 ohms.  This could be carried on indefinitely  but you get the point.  A little ingenuity and a soldering iron can take you a long way.  By the way, here's a commercial resistance substitution box.  It looks like something Tom Edison might have used, and probably is.

Fig 9-5

Now, for the bad news. Resistance and capacitance decades are great for audio frequencies and very low RF but, as frequency gets higher, the length of the attaching leads and the internal wiring becomes a significant portion of a wavelength. The result is, that if these devices work at all, they will introduce so much reactance into the circuit as to make them useless. For most RF circuits, you'll just have to solder substitute components in place with leads as short as possible.

## Attenuators

Often, you'll find it necessary to reduce the amount of signal from a generator to a value well below its minimum output  There are two ways of doing this.  One is by inserting an attenuator between the signal generator and the equipment being tested.  Here's a commercial attenuator that is selective in small increments from 1 to 30 dB.

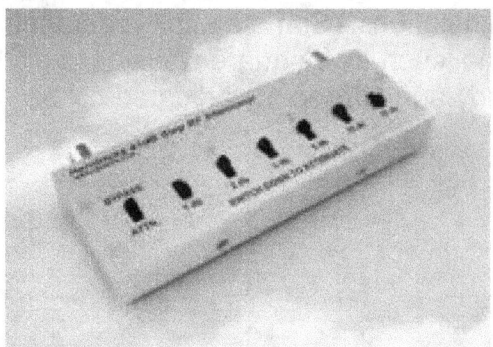

Fig 9-6

This is useful if you need to know the exact amount of attenuation.  If it is simply a matter of reducing the signal level to a vary low level, you can usually accomplish this by loosely coupling the signal generator to the equipment being tested.  Often this means just using a short piece of hookup wire as an antenna and laying it near a circuit.  Be careful to do this only with extremely low level signals.  I once tried to adjust a bandpass filter using this method.  I found an unshielded 1, watt, 2 meter signal on a five inch piece of wire, sufficient to trigger a repeater 30 miles away!

## Probes

This brings us to the subject of test probes.  Probes of all sorts are commonly used to insert a signal into a circuit, or to connect the input of an instrument to measure something.  One of the most common probes in use is the X10 oscilloscope probe -used to extend the measurement range by a factor of 10.  Here's a standard probe set for use with a *Tektronix* oscilloscope.

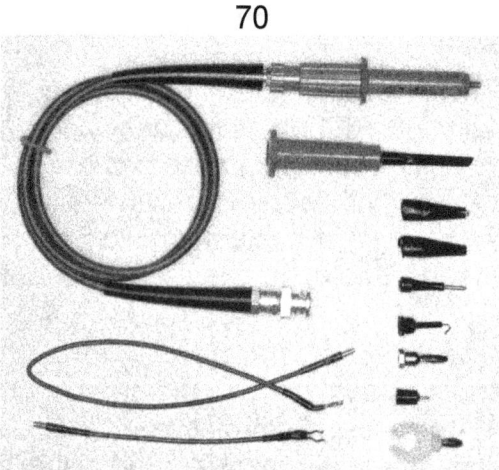

Fig 9-7

It consists of a length of coaxial cable, a probe body, and interchangeable tips to connect to the circuit being tested. All probes of this type are subject to voltage and frequency limitations. Some exhibit reactances that can be nulled by shaping a standard signal such as a square wave. Here's how a square wave signal looks for a properly and improperly adjusted probe

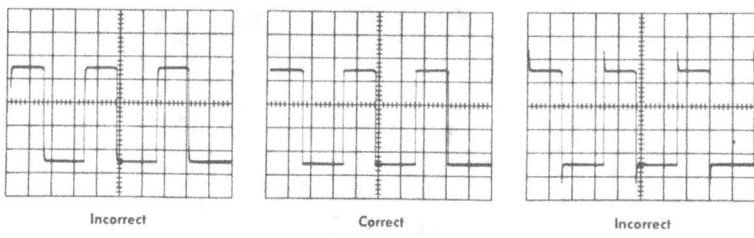

Fig 9-8

Probes needn't be as sophisticated as the Tektronix probe. They can be as simple as a piece of hookup wire bent to a particular shape, or they can contain multiple elements in a complex assembly. Here's a compensated RF probe used to calibrate an *Elecraft* K2 transceiver. It contains a resistor, capacitor and rectifying diode. mounted on a sliver of circuit card. The probe is a short length of bare copper wire. Cheap but effective!

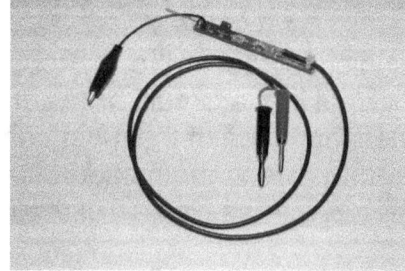

Fig 9-9

Another type of probe I find useful is a home-brew RF pickup loop. It's a simple air-wound 1:3 transformer in a shielded box that you can insert between a transmitter output and a dummy load. It couples an RF signal to a frequency counter, o-scope, or spectrum analyzer as needed.

Fig 9-10

As you can see form this internal view, the transformer primary is a piece of 16 gage bare copper wire with a 1" loop, centered between type N coaxial connectors. 3 loops of hookup wire connected to a third type N connector, comprise the transformer secondary. You can refer back to Fig 5-11 to see how it is used to monitor a transmitter output signal.

Fig 9-11

## RF Power meter

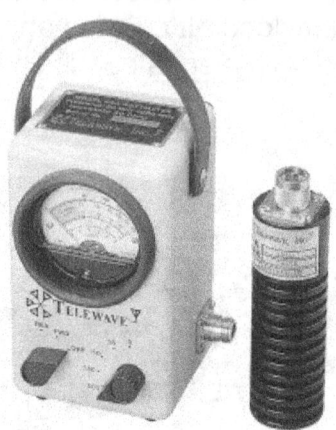

Fig 9-12

As I mentioned before, you can tune your transmitter for maximum output using a simple fluorescent or neon lamp but it won't tell you how much power you're actually putting on the air.  If you don't know, you could be in violation of FCC rules or the safety of anyone in the proximity of your transmitter or antenna. You might also be interested in how much reflected power you're getting back because of mismatches in the antenna system.  Both these measurements can be made by detecting a calibrated amount of RF power, demodulating it, and displaying it on an analog or digital meter.  Fig 9-12 shows a rather expensive but very accurate RF wattmeter.  It is a laboratory instrument, useful for precise transmitter calibration.  More likely, you'll want something that can be inserted in series with your antenna system.  Perhaps it might even be a permanent installation.  Fig 9-13 shows how these measurements can be made.

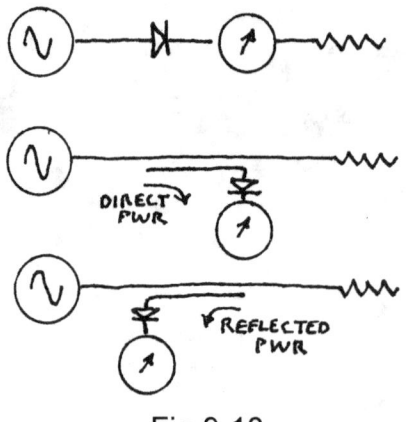

Fig 9-13

The top diagram shows the meter being used to calibrate the power output of a transmitter.  It is operating into a dummy load of the same impedance as an antenna system.  The circuit in the middle diagram uses a directional coupler to sample the forward power into a load.  In this case the load is most likely to be an antenna.  The lower diagram shows the same scheme to measure reflected power or standing wave ratio (SWR).

A typical metering system used by ham radio operators to monitor both forward and reflected power as shown in Fig 9-14.  A bi-directional coupler handles both direct and reflected power.

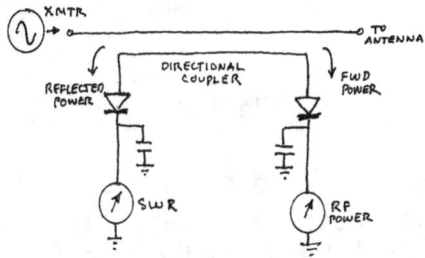

Fig 9-14

To measure both signals simultaneously, a special "cross needle" meter is used. (Fig 9-14)  The left hand needle indicates direct power to the antenna and the right hand needle indicates the reflected power.

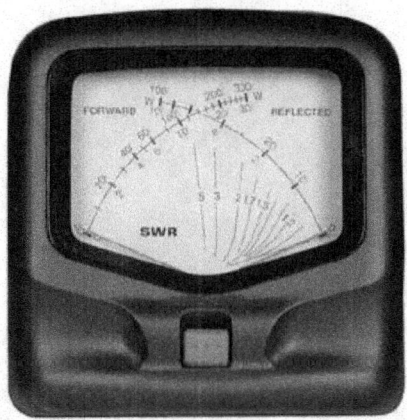

Fig 9-15

Figure 9-16 is an example of how a single meter is switched between the forward and reflected power monitors.  It still uses a bi-directional coupler but in addition to a simple reflected power reading, it also reads SWR, peak and average power in three ranges: 5W, 20W, and 200W peak.

Fig 9-16

This completes the overview of test instruments. It is not complete. It would be impossible to include the thousands of specific types and variations of instruments available but I think it includes the most frequently used instruments, at least the ones I have used for the past 70 years.

## Answers To Exercises

### Unit 01
01-1) R=E/I,  R=10/0.05,  R200Ω

01-2) I SHUNT =0.95A        R=E/I,  R=10/0.95,  R SHUNT= 10.52Ω

01-3) R TOTAL =100/0.050,  R TOTAL =2000Ω,  2000-200=1800Ω

01-4 24X10=240Ω

### Unit 02
02-1) 25.00

02-2) Maximum = 25.5+0.255 (+3)= 25.758.  Display =25.79

　　　Minimum = 25.5-0.255 (+3)= 25.248.  Display = 25.25

### Unit 04
04-1) 0.001x10=0.010   1/0.010=100Hz

04-2) 0.1X3.25=0.325V P/P  0.325/2-=0.1625V peak

### Unit-07
07-1) 1.499 to 1.501

07-2) 1.500

## *About the author*

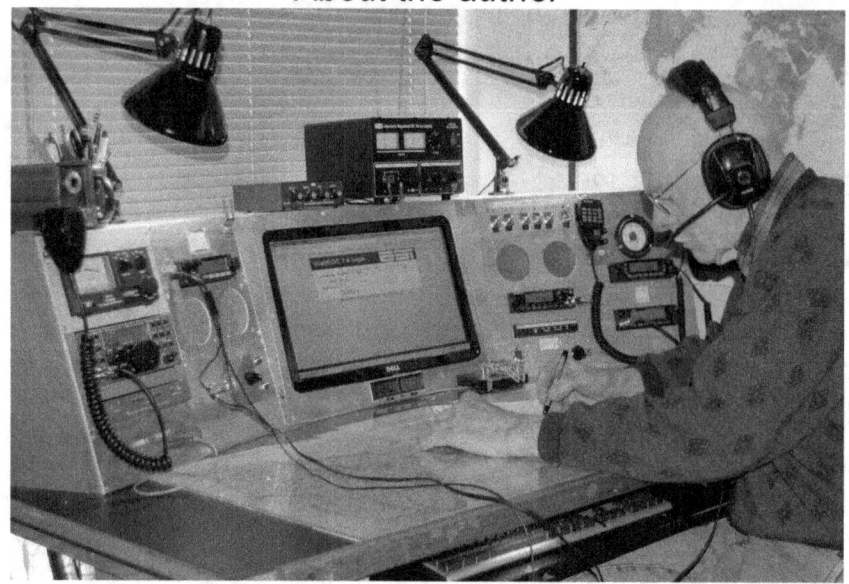

For many years, a varied career and even more varied interests took me away from the electronic field. I even tried to ignore a previous brush with ham radio as a student in high school. "Like riding a bicycle, once learned, never forgotten." Years of home ownership, appliance maintenance, troubleshooting electrical faults, rewiring stuff, kept the old skills sharpened without my realizing it.

When I renewed my interest in ham radio to become qualified as an emergency communicator, I never intended to get involved in equipment construction and maintenance again. So much for intentions. What started innocently, as a commitment to teach some classes in basic electronics, soon acquired a life of its own. Here I am, up to my ears in hardware again, teaching, writing, and enjoying it thoroughly. Technology is evolving rapidly and there's so much to learn and to experiment with. How can anyone watch these new developments without a sense of excitement? Ham radio is a means to get involved, an opportunity to help shape the future of communication. Grab hold and let it take you where it will. It promises to be a hell of a ride!

By the way, In my book, *The Practical Radio Shack*, I showed how my own station had evolved over the years and hinted that the process was not yet complete. The picture above is just the latest incarnation. I expect it's still not complete.

Paul Honore' W6IAM